Comprehensive™

BIOTECHNOLOGY 4

Comprehensive™

BIOTECHNOLOGY 4

Including Molecular Biology

S. Mahesh

Pushpa Agrawal
Department of Biotechnology
R.V. College of Engineering, Bangalore

PUBLISHING FOR ONE WORLD

NEW AGE INTERNATIONAL (P) LIMITED, PUBLISHERS
(formerly Wiley Eastern Limited)
New Delhi • Bangalore • Chennai • Cochin • Guwahati • Hyderabad
Jalandhar • Kolkata • Lucknow • Mumbai • Ranchi
Visit us at www.newagepublishers.com

Published by New Age International (P) Ltd., Publishers
First Edition : 2006

ISBN : 81-224-1828-7

Rs. 95.00

C-06-01-548

Printed in India at Hindustan Offset, Delhi.
Typeset at Shubham Graphics, Delhi.

PUBLISHING FOR ONE WORLD

NEW AGE INTERNATIONAL (P) LIMITED, PUBLISHERS
(formerly Wiley Eastern Limited)
4835/24, Ansari Road, Daryaganj, New Delhi - 110002
Visit us at **www.newagepublishers.com**

Preface

Biotechnology has emerged as exciting area in recent days. The impact of biotechnology on the quality of mankind is exemplary. It has acquired distinct academic description in recent years.

Biotechnology is one of the fastest emerging fields as it comprises of multidisciplinary subjects like medicine, agriculture and environment. Several foundational subjects, namely, molecular biology and others have strengthened biotechnology and contributed significantly in developmental process. Understanding the basic functions of biomolecules is indispensable and can act as a key tool in developing new techniques in several emerging areas of biotechnology.

The present book covers the syllabus of second year B.Sc degree course in biotechnology (fourth semester) prescribed by Bangalore University. This book is also useful for the students studying B.E. course in biotechnology prescribed by Visweswarayya Technological University (VTU). This book endeavours to furnish a simple, understandable text for students. This book contains eleven chapters on molecular biology. Sufficient illustrations are given wherever possible.

We express our sincere thanks to Saumya Gupta, Managing Director and V.R. Babu, Manager Marketing, New Age International Publishers, New Delhi, for their kind acceptance in publishing this book.

S. Mahesh

Pushpa Agrawal

Contents

1

CHAPTER

Molecular Basis of Life

1. INTRODUCTION

One of the most outstanding contribution in the field of biological scence is the discovery of DNA structure and the role of RNA displayed in the synthesis of protein. The spectacular success in the analysis and comprohensive understanding of zenetic code and the nature of zene regulation laid foundation in the field of molecular biology. This molecular approxes in turn have shown greater diagram of positive inluence in understanding traditional disciplines such as genetics, cytolosy, embryolosy and evolutionery biology.

Several refined work pertaining to molecular biology is based on the work carried out by bacterial experimently. One of the most simple and reliable organisor in strong thring the studis is Eschericahic col:. Molecular biology is not only pertaining in understanding of biomolecular of prokaryotes but also eukaryotic cell. The fine tuning in understanding the cell function of higher organiors laid a foundation for the inventing stratesis of genetic manupulation of cells and future cutting edge technology.

The primitive earth is believed to have formed about 4,800 million years back, as a hot fire ball without atmosphere. The cooling process of earth might have taken place for millions of years which resulted in random collision between neutrons and protons to form atomic nuclei. Further cooling resulted in binding of electrons to atomic nuclei. When accumulated gas inside the earth escaped slowly from the earth's crust through surface cracks it formed atmosphere. Water was formed immediately and entered atmosphere due to evaporation. As the water recycled, continuous oceans were formed as a result of torrential rains for millions of years. In the beginning there was virtually no atmospheric oxygen on earth. The present day life entirely depends on the availability of oxygen on this planet which was originated millions of years back.

The nature and types of atmosphere on primitive earth was debated for a long time. Undoubtedly primitive earth comprised of a set of gases like hydrogen, methane and ammonia. Hydrogen sulphide and carbondioxide were probably present permanently as an essential component of primitive earth. The available condition on primitive earth promoted several chemical reactions to take place in the atmosphere as well as in the gases that dissolved in water. The energy for these reactions were derived from the sun radiation as well as from the electric discharge. Further, accumulated gas and volcanic eruptions discharged gases into the atmosphere.

The life is originated from primitive molecules million years back. However, there are fundamental questions on how did biomolecules arise? Biomolecules perhaps started interacting with each other and organised into specific function. It was amazing to presume how the first cell originated from these well organised biomolecules. Several decades of fundamental studies reveal adequate information on the possibilities of generating biomolecules under laboratory conditions.

1.1 ORIGIN OF ORGANIC MOLECULES

Primitive earth consisted of wider variety of important biomolecules at sufficient concentration and acted as building blocks to construct primitive life on this planet. The spontaneous prebiotic evolution suggested this view. During 1920 and 1930 considerable work has been done by J.B.S. Haldine and A.I. Oparin on the process of spontaneous production of nucleotides and primitive polymers.

Several pieces of evidence suggested that several organic molecules might have synthesized originally from abiotic origin on the primitive earth (primeval earth). The non-oxidizing earth's atmosphere undoubtedly contained H_2S, NH_3, CO_2, CH_4, H_2 and N_2 except oxygen. The presence of sufficient concentration of precursor molecules promotes the synthesis of organic molecules using sun's radiant energy and thermal energy provided by volcanic eruptions. The hydrogen was soon lost as a result of oxidised component. The oxygen appeared much later and its concentration built up gradually due to continuous photosynthetic activity on the earth.

1.1.1 Miller's Experiment

Stanely Miller, a student of Harold urey, conducted series of interesting experiments in 1953, to demonstrate the synthesis of organic compounds under invitro conditions, in which he simulated possible prebiotic conditions prevalent on the primeval earth. He mixed methane, hydrogen and ammonia with water in a closed reflux vesel. Inside the reflux vessel the mixed gases which simulated the early atmosphere was exposed to electric disharges. The simulated primeval ocean contains more than 10 percent of the carbon (from methane) was included in the organic molecules like glycine, alanine, valine and leucine. Several intermediate products like hydrogen cyanide (HCN), aldehyde and cyano compounds were produced in the invitro experiments. These intermediates could have been an important intermediate in the formation of amino acids and nucleic acids. Miller's experiments have demonstrated different reactions that mimic prebiotic conditions that are responsible for the synthesis of different aminoacids. He mixed hydrogen cyanide, aldehyde and cyano compounds with H_2S and supplied ultraviolet light to provide energy for the reactions which lead to the formation of different aminoacids.

Invitro simulated experiments showed that reactions can be initiated by thermal energy released by volcanic eruptions and ultraviolet from solar radiation and electric discharge from lightening under simulated condition. Hydrogen cyanide in turn is the precursor for other reactive compounds such as cyanamide, cyanoacetylene and nitrites. These further act as fine precursors of various aminoacids, purine and pyrimidine. Heating the mixture of ammonia and hydrogen cyanide under simulated conditions of primeval earth results in the production of number of amino acids and considerable amount of purine adenine.

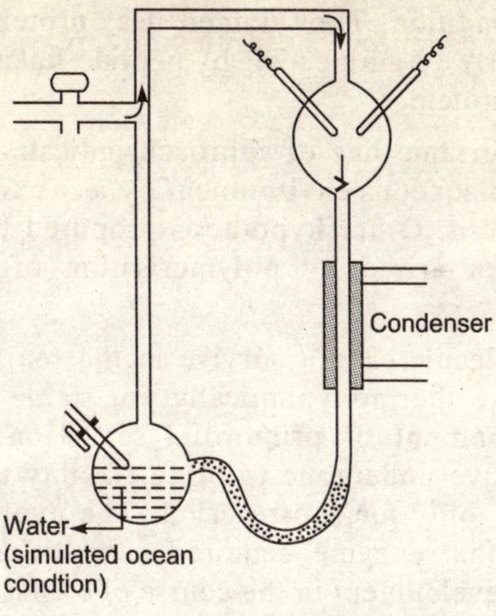

Fig. 1.1 *Miller's experiment.*

Ultraviolet and lightening discharge were the two important sources available on the primitive earth that have been found to activate the components of the primitive atmosphere to generate organic molecules.

1.2 PREBIOTIC ORIGIN OF COMPLEX BIOMOLECULES

When primordial building blocks were formed from abiotic synthesis, further chemical organisation resulted in complex biomolecules. This must have been achieved by covalent bond linking the building blocks to form more complex molecules such as polysaccharides, polypeptides and polynucleotides. The covalent bond linkages between the building blocks can take place by condensation reactions which involves removal of water molecules. However, some apprehensions were raised about these linkages due to thermodynamically unstable condition of peptide, ester linkages and glycosidic bonds in aqueous condition. Hence, possibilities of small amount of such linkages can exist. Many speculations were followed in the process of condensation reactions. One is the involvement of chemicals in condensation reaction which was believed to extinct water molecules in the reaction facilitated covalent linkages. However, S.W. Fox, a leading investigator on origin of life threw light on the exact nature of chemical reaction. Covalent linkages which binds prebiotic polymers were largely thermal in nature. This presumption was evidenced when a polypeptide was formed by heating the mixture of aminoacids at about 150°C. Among all these reactions most relevant biological condensation reaction can take place by condensing agents such as polyphosphates and polyphosphate esters which can bring dehydration of peptides from aminoacids on irradiation with ultraviolet or heating.

1.3 THE PREBIOTIC ORIGIN OF POLYPEPTIDES

Polypeptides are proteins which are synthesized by polymerization of amino acids. The experiments conducted by Fox and his colleagues have shown that polymers resembling polypeptides were formed when a mixture of aminoacids are heated or subjected to electric discharge or when they were treated with condensing agent like polyphosphates under simulated primitive earth condition. They named it as proteinoids which approach 20,000 daltons. They comprised nearly 18 aminoacids by peptide linkage. They exhibit several other properties characteristics of protein.

Further experiments in the same line of approach indicate that proteinoids may even be formed at low temperature in aqueous environment by a reaction similar to aminoacyl-tRNA in present day protein synthesis. Other hypothesis proposed by Akabori has suggested that polypeptides may have been arised by polymerisation of monomeric units instead of aminoacids.

Any single proteinoid molecule cannot survive in the sea for long time due to unstable peptide bond, since these are thermodynamically not stable and may undergo hydrolytic cleavage in aqueous environment of primordial sea. Hence, improvisation process of aminoacid sequence could have undergone to bring stability to the proteinoid biomolecule. They can be able to survive only for short period. The evolution of enzyme from simple precursors seems probable that enzyme acquired catalytic activity first and followed by substrate specificity as later development in the course of evolution. The first primitive catalyst could have been a simple short polypeptide consisting of α–ammonium group, an imidazole ring or a carboxyl group. Further selection and modification of polypeptides may have evolved and formed more potent peptide with more specific catalytic activity. In addition, they were stable and equipped with better survival in the primitive primordial environment. The earliest proteinoids may have been the precursors of enzymes. Several rate enhancing activities have been detected in proteinoids like hydrolysis of esters, decarboxylation of α-ketoacids etc. The progress made in the evolution of catalyst was meagre until the availability of long peptide chain which was capable of assuming three dimensional status.

1.4 ORIGIN OF NUCLEIC ACIDS

Several simulated primitive earth experiments were conducted to consolidate information on evolution of nucleic acid molecule. Nucleoside like adenosine and deoxy adenosine have been detected in simulated earth experiment. When nucleosides and polyphosphates are heated or irradiated with UV, nucleotides are formed. Using similar experiment ADP and ATP were generated. Next course of evolution is the evolution of nucleic acids by interlinkages of nucleotides. This was also evidenced by simulated earth experiment in which nucleotides were heated to 60–65°C in the presence of polyphosphoric acid as condensing agent. In the primordial period first polynucleotide conains 2′–5′ linkage. However, 3′–5′ linkages may have arisen due to availability of specific enzymes which have the ability to enhance 3′–5′ linkages of nucleotides.

Since nucleic acids are capable of storing the genetic material and are perfectly suited for the task of replication, it is difficult to image how the first nucleic acid genetic material could be replicated with the participation of protein catalyst. Ambiguity continues as how a protein catalyst could arise without a nucleic acid genome which encodes it. Several classical theories were proposed in 1981 to overcome ambiguity in the first origin of nucleic acids. It was shown that rRNA introns could function as enzyme suggest that very first living molecules might have been an RNA replicase. It can be even suggested that this enzyme catalysed its own replication without help from a protein. This hypothetical RNA might have been behaved functionally as both the genetic material and replication enzyme. In the due course of investigation many new examples of catalytic RNA have been discovered. Hence, catalytic features of RNA suggest that they are molecular fossils. Studies of catalytic RNA threw light on the current views about the origin of life. The origin of the replicase enzyme can be linked to random prebiotic condensation of mononucleotides which later formed longer and longer polymers. This earliest or first RNA replicase would have been able to use either itself or another nucleic acid as a template for polymerization of nucleic acid precursor found in prebiotic soup.

The first origin of biomolecules like proteins and nucleic acids is under debate. Many are of the view that primitive proteins originated first. The first protocell formed when primitive catalyst was entrapped in gel like matrix or first surrounded by a membrane. During the process some how acquired a primitive metabolism. In contrast it was presumed that first cell functioned in the absence of nucleic acids away from the genetic system, which they acquired in the later stage of evolution. Second approach is that nucleic acid formed first and acquired the genetic information for the evolution of proteins. Third approach is that both the proteins and nucleic acids come together to a closer proximity and evolved to form a first primitive cell. This can be explained on the basis of following hypothesis.

According to Gamow in 1959, there are contrasting differences between aminoacid sequence and nucleic acids, perhaps due to differences in their physiochemical characteristics. The evolution of polypeptide and polnucleotides might have probably occurred simultaneously. It has been postulated that polypeptide with random assembly of aminoacid to possess some catalytic activity.

1.5 ORIGIN OF PROTOBIONT

Protobiont considered as primitive forms of life, might have been originated first in the primeval sea. In order to consolidate this theory, several evolutionary investigations were carried out and proposed some models. They are coacervative model and protenoid model.

1.5.1 Coacervate Model

The first precursors of cells called protobionts arose when membranes formed around macromolecules exhibiting catalytic activity such as proteins. Oparin suggests that their cell phase could have arose from the concentrated primordial broth by the process of coacervation. Thus coacervative droplet that was formed spontaneously when water molecules are organised around electrically charged molecule. Some of which even fused to form complex coacervate. They are able to concentrate organic molecules from their surroundings. Since they conain enzymes which are able to synthesise and degrade many complex molecules such as starch, release maltose into surrounding medium. In the simulated experiment concentrated solution of polypeptides, polysaccharides and even RNA could be induced to form coacervative droplets of different dimensions to which enzymes can be easily incorporated into such droplets and several metabolic activities can be demonstrated. Oparin's coacervative droplets are simply a model and they do not exhibit any self replication process or participate in evolution.

1.5.2 Proteinoids

According to Sydney Fox in 1965, one of the most favourite choice for protobiont is proteinoid. They are protein-like molecules synthesized under invitro condition by polymerization of aminoacids when heated between 130°–160°C. Proteinoids maintain stability between 0–25°C but are unstable at high temperature. In addition, they exhibit weak catalytic properties. It was also found that proteinoids were found to possess an outer membrane boundary resembling plasma-membrane structure which were devoid of lipids.

1.6 ORIGIN OF GENETIC CODE

Several investigators are of the opinion that code on the DNA molecules arose gradually in stages and not formed at once for all. In the evolutionary selection, force for the development of the code must have come from the primitive protein. Primitive polypeptide must have comprised mainly of glycine and other aminoacids which have organised into useful oligopeptide sequence carrying out some primitive functions. Since RNA molecules was postulated as earliest molecules, information in RNA molecules is encoded in the form of triplet codes. But it was a matter of debate on the evolution of this code. Among existing codes those specifying the same aminoacids have common first two bases. However, the third base is different. The existing two base code could code for only 16 types of aminoacids. It was postulated that functions of primitive protein might have produced by the two base genetic code. Due to instability of this doublet codon mode, they could not have stabilized for long time. As a result triplet codon was perhaps formed during the course of natural selection.

According to Franscis Crick in 1967, organisation of genetic code was the ultimate result of random interaction over time and space. Finally development of genetic code was stabilized. Any further change in the codon usage during evolution was restricted due to lethal effect that would have inflicted, if any disturbance to the aminoacid sequence.

1.7 BIOCHEMICAL EVOLUTION AND PROKARYOTIC ORIGIN

One of the most simple living organism is mycoplasmas which exhibit many similarities. Mycoplasmas measures 0.3 μm and contain nucleic acid code for about 750 types of proteins. Survival of cell depending on the minimum number of proteins. Both RNA and DNA shows genetic information. Once the primitive cell was evolved, due to competition with others it acquires its own new metabolic engineering status from the surrounding. Due to continuous interaction and rapid changes in the environment mycoplasmas like cells evolved themselves into prokaryotes such as bacteria and blue green algae.

Once the prokaryotes was evolved they were exposed to wide range of shuffled environmental condition. Availability of different organic molecules in the environment compel the bacterial orgaisms to reorient or evolve a new metabolic pathways to effectively utilise the available food material. Thus a new set of enzymes are required for the sustaining of newly evolved pathways. Hence elaborate metabolic pathways might have developed.

In the primitive earth conditions, for the first evolved life, there was little need for elaborate metabolic pathways and were able to sustain on the primordial soup or organic molecules. Availability of natural resources dwindled considerably as the population of cells increased to the greater extent. Due to competition for food even a simple bacteria had to evolve a new line of metabolic pathway in order to utilise more organic food material in the environment. Due to increase in metabolic related enzyme and further cellular adaptation it was suggested that diverse metabolic pathways have invariably evolved in the course of evolution.

1.8 ORIGIN OF EUKARYOTIC SYSTEM

Eukaryotes evolved from large anaerobic prokaryotic cell. Due to increase in demand for energy prokaryotic entities may have gradually transformed into eukaryotic mode of life. Such anaerobes may have engulfed small photosynthetic or heterotrophic prokaryotes, which later became metabolic endosymbionts and are well equipped to furnish glycolytic fermentation and respiratory capacity. It was presumed that endosymbionts may have been the key players for the formation of modern chloroplast and mitochondria.

Establishment of mitochondrial organelle system in eukaryotes provided a significant boost in generating energy which greatly enhanced their survival and adaptation potential besides better equipped to face the environment. They have acquired several modes of differentiation by modifying their own genome. The course of biological evolution is mainly due to the occurrence of non lethai mutation which is the result of chemical or physically induced changes in the base sequence of the genome. According to Sydney Fox efficiency of primitive photosynthesis system was increased by the development of membrane bound lamellae as in the case of cyanobacteria. Emerging oxidised atmosphere stimulated primitive photosynthesizing cell to break up internal photosynthetic membrane in the small sacs. During further course of evolution some cells acquired different version of these membrane sacs. Some of the cells are equipped with well organised chloroplast. Whereas other cell group began to lose their photosynthetic process but retain their electron transport chain and transformed into mitochondria. Normal cells required one set of chromosome. When the genome are subjected for deleterious mutation they were perished in the long term. While others with non lethal and favourable mutation survived besides adapting to new condition.

According to Smith in 1978, nucleus is formed by invagination of the surface membrane which make the boundary surrounding the DNA molecule, which provides compartmentalization to separate genetic material from the rest of the cytoplasm for the formation of nucleus, and circular DNA increases its size to acquire more genetic complexity. Further suggestion reveals that they break up linear DNA into pieces which result into line of division to increase its copy number. Further biochemical evolution evolved lyosomes by enzyme containing vesicles. Besides formation of golgi apparatus and endoplasmic reticulum by mesosomal investigation and its further extension into the cytoplasm as a tubular network.

One of the most significant event in the course of evolution is the appearance of mitochondria and chloroplast in plant cells. The genetic material of these cell organelles especially mitochondrial DNA resembles prokaryotes in posessing circular DNA and reproducing by fission. However, this high energy producing power house organelle favours the eukaryotic system for well adaptation. In the similar fashion chloroplast too share some of the above common features with prokaryotes. Presence of contrasting similarities between eukaryotic photosynthetic organism with cyanobacteria is pertaining to photosynthetic activity. Thus, origin of chloroplast is presumably from ancestral cyanobacteria, which at some stages of evolution was engulfed by eukaryotes. It was evidenced by the agrobacteria like structure of the chloroplast present in the present day moss cells and further evidenced is the occurrence of cyanobacteria in certain present day eukaryotic cells such as cyanophora. Due to continuous organisation of genetic complexity there is considerable increase in gene dosage and well characterized differential gene regulation make the eukaryotic system as well stabilised organism.

1.9 EVOLUTION OF AUTROTROPIC MODE OF LIFE (*PHOTOSYNTHESIS*)

The primitive cell in the primeval period were heterotrophic and also exhibited anaerobic mode of life. They used to procure nutrition from the organic soup. In the course of evolution, as the nutrition became scarce due to increasing population of cells, several organisms invariably acquired different mode of nutritional pathway. Some of the organisms synthesized this food material utilising sunlight and green chlorophyll. These are called phototrophs. Origin of chlorophyll in these autotrophs is one of the turning point in acquiring photosynthesis. However, majority of the bacteria derive their energy from inorganic molecules and synthesize their own food material at photo-synthetically.

In the evolutionary adaptation primitive cell modify its genome so that complex array of enzymes can be produced to utilise sun's energy more efficiently for the synthesis of complex organic food material. In comparison phototrophic bacteria acquired organisational complexity and wide array of enzymes and make the bacteria, a challenging task of reducing carbon dioxide into the complex carbohydrate. The imposing selection pressure makes the phototrophs to reorient themselves into a difficult mode of photosynthetic activity in which water was electron donor in some cells whereas other cells developed a system to release sufficient amount of molecular oxygen into air and water. While some of the cells developed respiratory electron transport chain in which molecular oxygen is the final electron acceptor. In the competition the light senisitive chlorophyll pigments must have evolved to utilise solar energy in photons in competition with earlier process of photophosphorylation of nucleotide to form high energy compounds like ATP.

SUMMARY

Primitive earth was formed millions of years back as a hot fire ball devoid of atmosphere. Atoms were formed due to continuous collision between neutrons and protons then joins electron to become atom. Gaseous molecules and water appeared due to torrential rains. Due to several primeval chemical reactions which took place with the help of energy obained from electric discharge of lightening and sun radiation, primitive earth consisted of wider variety of important bimolecules which acted as building block to construct primitive life on this planet.

The primitive atmosphere of the primeval earth contained mainly carbon dioxide, methane, sulphur dioxide, ammonia without oxygen. Hydrogen was lost due to oxidised component and oxygen appeared much later due to photosynthetic activity. Stanely Miller conducted several experiments under simulated primeval earth conditions. He had synthesized organic compounds under invitro conditions by mixing methane, hydrogen and ammonia in presence of water. In addition several intermediate products were produced in invitro experiments like hydrogen cyanide, aldehyde etc. These intermediate were important molecules in the formation of aminoacids and nucleic acids.

Several chemical organization resulted in complex biomolecules by possible covalent linkages. The building blocks forms more complex molecules such as polysaccharides, polypeptides and polynucleotides. Several biological condensation reactions took place by condensing agents such as polyphosphates and polyphosphate esters which can bring dehydration of peptides from aminoacids. Several simulated experiments were conducte to evidence origin of polypeptides. The experiments include the heating of mixture of aminoac ds or were exposed to electric discharge or to condensing agent like polyphosphates and obtained polymer resembling polypeptides known as proteinoids. Due to unstable nature of protenoids molecule they undergo hydrolytic cleavage. Hence, improvisation of protenoids aminoacids could have undergone to bring stability to the protenoid biomolecule. Then enzymes were evolved from simple precursors. It seems probably that enzyme acquired catalytic activity first and followed by substrate specificity as later development in the course of evolution.

Additional simulated experiments were conducted to consolidate evidence on evolution of nucleic acid molecule. In the invitro experiments nucleosides and polyphosphates were heated irradiated with UV as a result of which nucleotides were formed. In the next course of evolution is the evolution of nucleic acid by interlinkages of nucleotides. There was ambiguity on how the first nucleic acid was able to replicate and how a protein catalyst could arise

without a nucleic acid genome which encode it. It was evidenced that first living molecules might have been a RNA replicase and are responsible for its own replication. This RNA behaved as genetic material as well as replication enzyme. Origin of replicase enzyme can be linked to the formation of long polymers. Several evidences proclaim that primitive proteins originated first and the first cell functioned in the absence of nucleic acids. Another view consolidates that nucleic acid formed first and acquired the genetic information for the evolution of proteins. Another approach claims that both protein and nucleic acid come to closer proximity and evolved to form primitive cell.

Protobiont was considered as primitive form of life and acts as precursors of cells. They could have arose from the concentrated primordial broth by the process of coacervation. The complex coacervates are formed in the later stage which acquired enzymes capable of degrading several molecules. Investigations are of the opinion that code on the DNA molecules arose gradually. Codes must have come from the primitive protein. First triplet code might have been in RNA molecule since it formed first. Unstable doublet codon perished, immediately as a result triplet codons were formed as a natural selection.

After the origin of prokayotes they were exposed to wide range environmental conditions. As a result bacteria were able to acquires new metabolic pathways and reoriented themselves to utilise new food material in the environment. Hence an elaborate metabolic pathway might have developed. During the course of evolution there was a serial transformation from the earliest heterotropic mode to autotropic mode of life. Some of the organisms started synthesizing their food material using sunlight and green pigments. These are called phototrophs. Chlorophyll origin was one of the turning point in acquiring photosynthesis.

Eukaryotes evolved from large anaerobic prokaryotic cell. As the requirement for energy demand increases considerably, prokaryotes might have transformed into eukaryotic mode of life. Due to engulfing of small prokaryotes the eukaryotes were able to acquire vital cell organelles during the course of evolution. Acquiring mitochondrial organelle by eukaryotes provides a significant boost in generating energy. One of the most significant event in the course of evolution is the appearance of mitochondria and chloroplast in plant cells. Organisation of complexity of the genome, differential gene regulation and stability of biomolecules are some of the highlights among eukaryotes as they as dominating all life forms on this planet.

QUESTIONS

1. Discuss the process of evolution of organic biomolecules.
2. Explain different concepts, and evidences in favour of molecular basis of life.

SHORT QUESTIONS

1. Coacervative mechanism
2. Miller's experiment
3. Organisation of prokaryotic cell
4. Evolution of eukaryotic cell

2

CHAPTER

Nucleic Acids

2.0 STRUCTURE OF DNA

DNA (Deoxyribonucleic acid) was discovered in 1869 by Friedrich Mischer. It is surprising to find that DNA, which has been the focus of work in Modern biology, was isolated almost a century ago. DNA is one of the most amazing biomolecule. It is present in all living organisms as linear molecule of extremely high molecular weight. Several group of scientists have shown their scientific vision on the nature of genetic material through their painstaking effort.

Although several scientific explorations have taken place in the field of unravelling of genetic material, the real story of molecular biology began in 1953 when two talented scientists James Watson and Francsis Crick crack down the structure of DNA. This was possible due to their painstaking analysis of X-ray diffraction pattern and by model building. This explosive invention persued closer look at the structure of the thread of life that carries all the hereditary information within the chromosomes.

2.1 DNA IS THE GENETIC MATERIAL

2.1.1 Evidence One

Several breathtaking investigations were carried out for considerable duration to prove that DNA is a genetic material. In one of the earliest classical experiment carried out by T. Avery and his team of researchers (1944) on bacterium Pneumococcus. The bacterium Pneumococcus kills mice by causing pneumonia, exist in different forms. Several types of Pneumococcus like I, II and III have different capsular polysaccharide. They have a (S) smooth appearance when smooth bacteria are killed by heat treatment they lose their virulent potential i.e., not able to kill animal. At several times it could be possible to see variants from smooth neumococcus bacteria. These variant bacteria have a rough (R) surface and are avirulent i.e., they do not kill mice due to the absence of polysaccharide.

In order to prove DNA as genetic material, first live smooth bacteria (virulent) was injected into mice, as expected, it died due to infection, In the second line of experiment when heat killed smooth bacteria was administered it could not kill the animal as it lost its virulance. In the third line of approach (experiment) they selected live rough bacteria and infected them into mice. It was found to be harmless. In the final stage of experiment, when a mixture of heat kelled smooth and live rough bacteria were injected the animal died immediately due to pneumococcal infection. Virulent S bacteria can be recovered from the dead mouse. The virulent bacteria recovered from the mixed infection had the smooth coat type of bacterium. The remarkable observation was that non capsular strain (rough bacteria) develop capsules proving that a chemical costituent in the extract of capsulated bacteria is potential of heritable change in the non capsulated avirulent strain. When extract was analysed, it was the DNA which brought about transformation. When Avery and coworkers isolated DNA from the virulent strain and mixed it with avirulent cells invitro transformation took place to mobile avirulent cells to virulent strain. Therefore it was proved that DNA is the active agent in carrying genetic information.

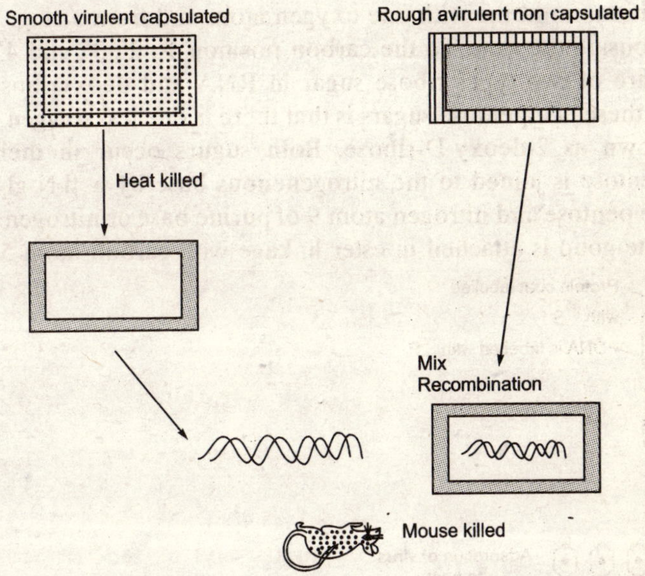

Fig. 2.1 *Avery experiment on pneumococcus transformation.*

2.1.2 Evidence Two

Although A.T. Avery and his team have made significant effort to label DNA as hereditary material by demonstrating the principles of tranformation, conclusive evidence was shown by Harshey and Chase in 1952 to strengthen the idea. These workers selected bacteria and bacteriophage virus as experimental organisms. Bacteriophages were labelled either in protein or DNA with ^{35}S of ^{32}P. The DNA could be labelled with ^{32}P as it contains phosphorus. The unlabelled bacteria are infected with phage with ^{32}P in DNA or with ^{35}S in protein. Adsorption of virgus to host cell takes place and labelled phage DNA entes inside the bacterial cell, leaving behind ghost coat protien. Phage coat contain 80 per cent of ^{35}S label an infected bacteria contain 70 per cent of ^{32}P. Bacteria when infected burst open after 20 minutes to release lange number of bacteria. These two are then separated by centrifugation. The bacteriophage labelled with ^{35}S did not exhibit radioacivity. However, when these were infected with ^{32}P labelled phages exhibit radioactivity. Further investigation reveals that when progenies of phage particles were isolated from infected bacteria and analyse. They contain only ^{32}P less than 1 per cent ^{35}S label. These experiments clearly strengthen the view that the packaged phage DNA enters the bacteria and labelled phage DNA becomes part of the progeny phages. Hence, it is the general conclusion that DNA is genetic material. Whether it is part of genome of cell or virus.

2.2 CHEMICAL COMPOSITION OF DNA

Purified DNA from all living orgaisms such as viruses, bacteria, plants and animals have shown that they are highly complex polymers made up of four monomers known as deoxyribonucleotides. DNA polymer consists of pentose sugar, a phosphate group and nitrogenous base. Nitrogenous bases are purines and pyrimidines. Purine bases are adenine and guanine and they are heterocyclic in nature. The pyrimidines are thymine and cytosine and are one ringed bases.

2.2.1 Pentose Sugar

The type of sugar present in the DNA is the pentose sugar consisting of five carbon atoms of which four carbon atoms combine with one oxygen atom and from a ring. The fifth atom forms-CH_2 groups reside ouside the ring. To the carbon position 1′, 2′, 3′, and 4′ hydrogen atoms are attached. Pentoses are of two types ribose sugar in RNA and deoxyribose in DNA. The only difference between these two pentose sugars is that there is one less oxygen atom in deoxyribose. Therefore it is known as 2-deoxy-D-ribose. Both sugurs occur in their furanose forms in nucleotides. The pentose is joined to the nitrogenenous base by a β-N-glycosyl bond between carbon atom 1 of the pentose and nitrogen atom 9 of purine base or nitrogen atom 1 of pyrimidine bases. The phosphate goup is attached in ester linkage with carbon atom 5 of the pentose.

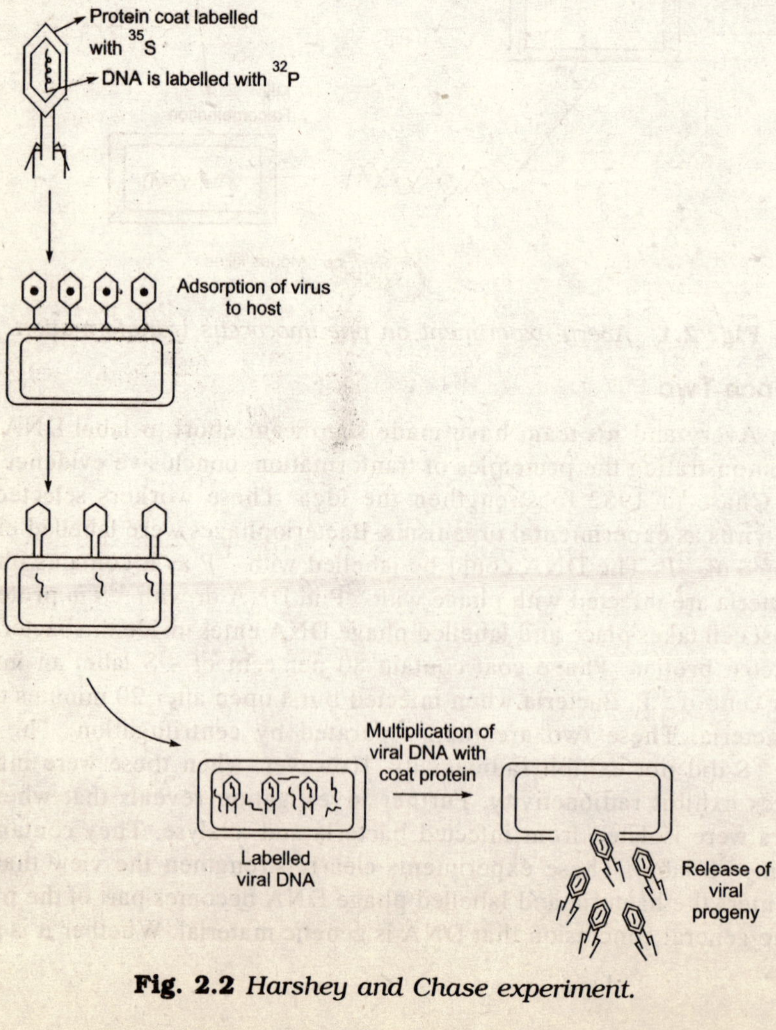

Fig. 2.2 *Harshey and Chase experiment.*

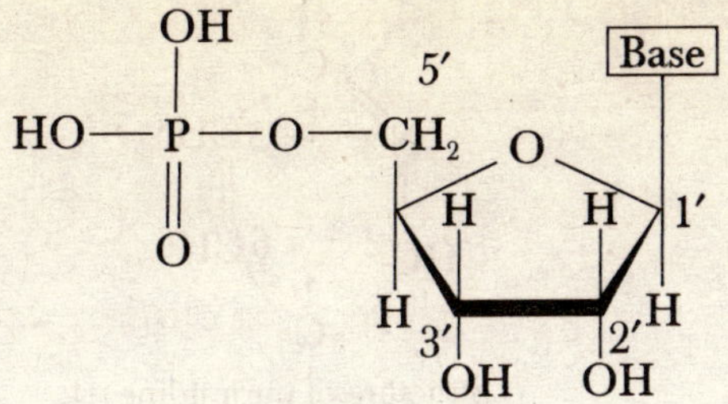

2′-Deoxyribonucleoside 5′-Monophosphate

2.2.2 A Phosphate Group

Strong negative charges of the nucleic acid is due to te presence of phosphate group. In DNA a phosphate group (PO_4^{-3}) is attached to the 3′ – carbon of deoxyribose sugar.

2. 2.3 The Pyrimidines and Purines (Nitrogenous Bases)

There are two classes of notrogenous bases, purines and pyrimidines. These two classes of nitrogenous bases found in nucleotides are the heterocyclic compounds. The purines are double ring compounds that basically consist of imadazole ring with nitrogen at 1′, 3′. 7′ and 9′ position. The pyrimidines are single ring compounds. They are planar molecules. In a single ring of pyrimidine the nitrogen being at position 1′ and 3′ in 6-membered ring. Pyrimidines are of three types Thyamine (T), Cytosine, Guanine and Cytosine are common in both DNA and RNA, However, Uracil is found only RNA.

2.2.4 Nucleosides

Nucleosides are the nitrogen base combined with pentose sugar. In DNA it is known as deoxyribonucleosides. Free nucleosides occur in small amount in most of the cells.

Necleoside = pentose sugar + nitrogenous base

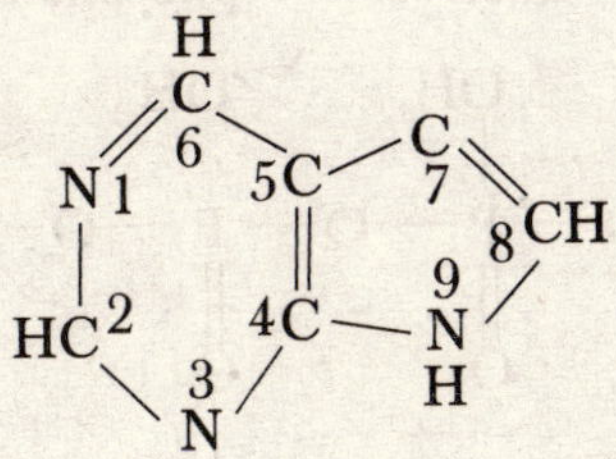

Structure of purine

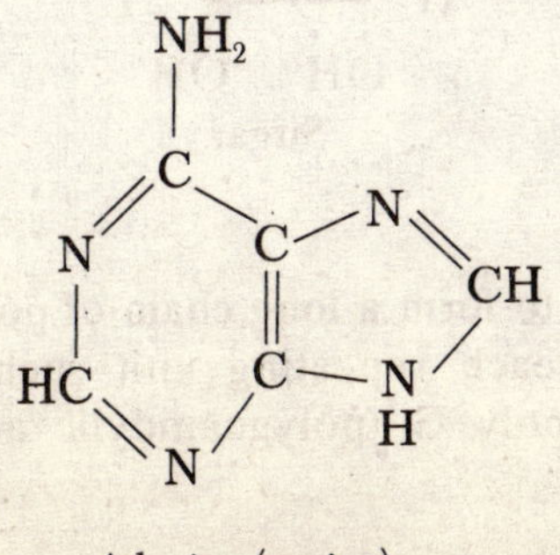

Adenine (purine)

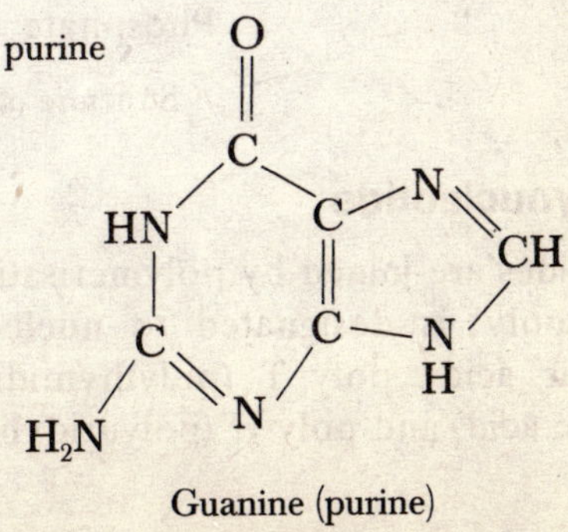

Guanine (purine)

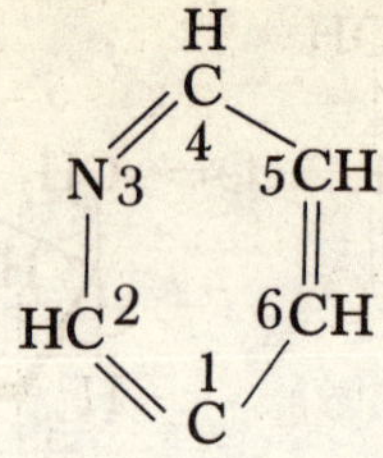

Structure of purimidine

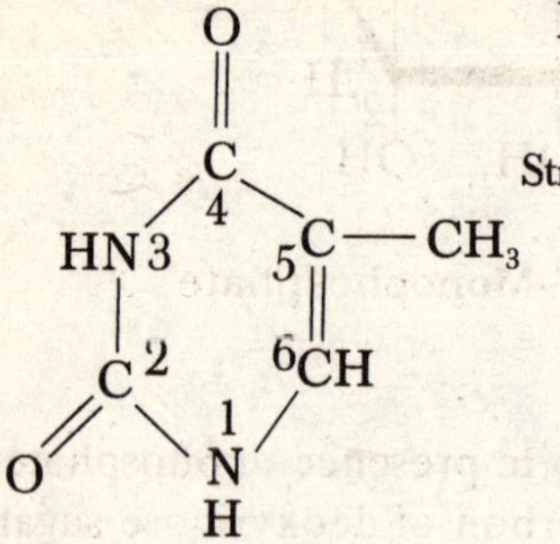

Thiamine (purimidine)

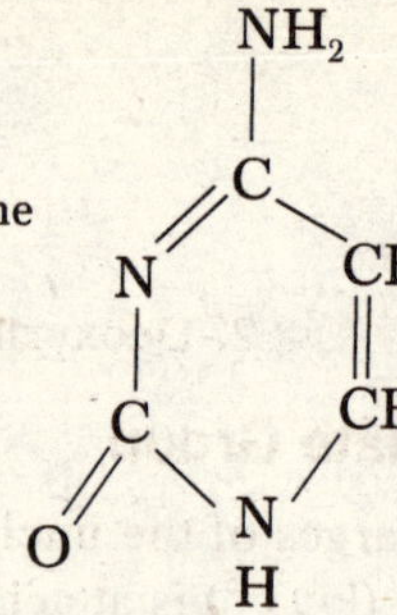

Cytosine (purimidine)

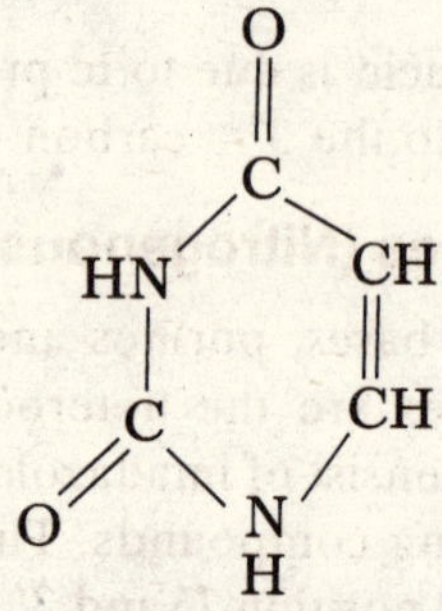

Uracil (purimidine)

2.2.5 Nucleotides

A nucleoside linked with phosphate forms a nucleotide. Deoxyribonucleotides occur in the free form in cells in significant amount. The free nucleotide forms availabe in cytoplasm are deoxyadenosine triphosphate (dATP), deoxyguanosine triphosphate (dGTP), deoxycytidine triphosphate (dCTP), and deoxythymidine triphosphate (dTTP).

Nucleotide = nucleoside + phosphate

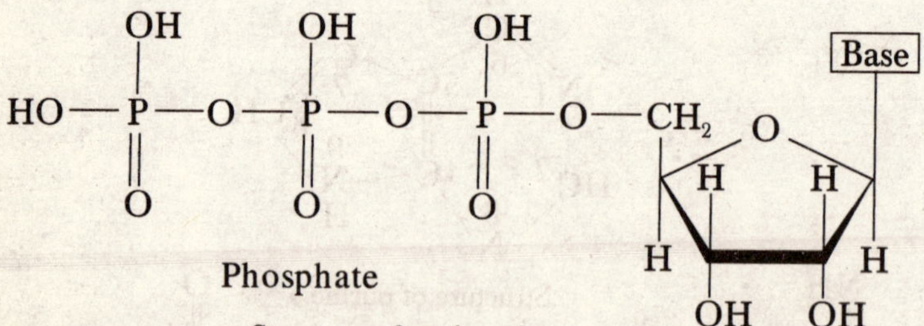

Structure of nucleotide

2.2.6 Polynucleotide

The nucleotides are joined by polymerisation process to form a long chain of polynucleotide. The term 'poly' is designated to nucleotides for each repeating unit such as poly A (polyadenylic acid), poly T (polythymidilic acid), poly G (polyguanidylic acid), poly C (polycytidilic acid) and poly U (polyuridylic acid).

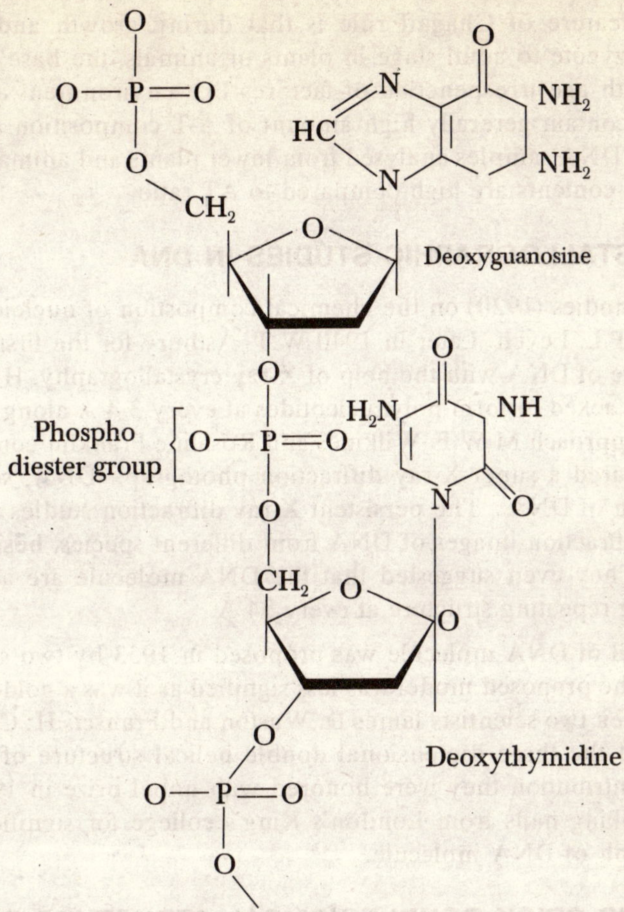

Structure of Polynucleotide chain.

2.3 DNA BASE COMPOSITION (*CHARGAFF-RULE*)

The genetic information of living organisms is stored and arrange in the form of a linear sequence of four bases (A, T, G and C). These four letter alphabet codes for the primary structure of proteins. According to previous evidence it was presumed that four major base composition of DNA occurs in equimolar ratio irrespective of any organism. However, between 1949 and 1953, a comprehensive study was carried out on base composition by Chargaff. According to these studies he found that although base composition varies from one species to another, in all cases the amount of adenine was equal to the amount of thiamine (A = T) and the number of guanine and cytosine was also found to be equal (G ≡ C). Further analysis showed that the total quantity of purines equals the total quantities of pyrimidines i.e., A + G = C + T. However, exact ratio of four nucleotide vary considerably between one species to another (AT/GC ratio). This is known as Chargaff rule. In addition, DNA of closely related species have similar base composition. When DNA was extracted from widely distant species, the analysed data reveals that there is more tendency to have different base composition.

Another salient feature of Chagaff rule is that during growth and development of an organism i.e., from zygote to adutl stage in plants or animals, the base composition of DNA does not change with age irrespenctive of factores like environment and nutrition, Higher plants and animals contain gererally high amount of A-T composition and low amount G-C content. In contrast, DNA samples analysed from lower plants and animals including microbes reveals that the G-C contents are high compared to AT ratio.

2.4 X-RAY CRYSTALLOGRAPHIC STUDIES IN DNA

One of the earliest studies (1920) on the chemical compostion of nucleic acid stems from the persistent aeffort of P.L. Leven. Later in 1940 W.T. Astbury for the first time proposed three dimensional structure of DNA with the help of X-ray crystallography. He highlighted that the flat nucleotide gets stacked to form polynucleotides at every 3-4 Å along the path of the stack. In the same line of approach M.W.F. Wilkines and Rosaline Franklin consolidated the work of Astbury. They prepared a super-X-ray diffraction photograph DNA, suggested evidence of helical configuration of DNA . The persistent X-ray diffraction studies revealed unequalness of identical X-ray diffraction images of DNA from different species, besides variation in their base composition. They even suggested that the DNA molecule are about 20 Å thick and 30,000Å long having repeating structure at every 34 Å.

Triple helix model of DNA molecule was proposed in 1953 by two scientists, Pauling and Convey. However, the proposed model was less signified as it was a golden time in the field of bilolgical science when two scientists James D. Waston and Franscis H. Crick from Cambridge University suggested the three dimensional double helical structure of DNA molecule. For their outstanding contribution they were honored with nobel prize in 1962. They shared the noble prize with wilking hails from London's King's college for significant studies of X-ray diffraction photograph of DNA molecule.

2.5 WATSON AND CRICK DOUBLE HELICAL STRUCTURE OF DNA

Advancement made in X-ray diffraction coupled with other techniques has made the greatest impact on the study of DNA structure. The X-ray diffraction techniques have been used extensively to analyse the structure of DNA by Franklin and M.H.F. Wilkins during 1950–53. Three dimensional structure of DNA was elucidated by X-ray diffraction measurement. Thanks to Watson and Crick (1953) for their persistent effort on constructing the model on the structure of DNA. According to this model it is composed of two right-handed helical polynucleotide chain that form a double helix around the same central axis. The double helix is right-handed in nature. The two strands of DNA are antiparallel i.e. their 3′–5′ phosphodiester links run in opposite direction. The coiling of two strands was well stabilized in such a way that they cannot be separated from each other. However, separation is possible only at the time of unwinding process. (Strand separation can also be achieved by heat denaturation of DNA.) Such a stable helical coiling system is known as plectonomic.

Double-helical chain is a polynucleotide, a polymeric collection of nucleotides in which the sugar of each nucleotide is linked by a phosphate group to sugar of adjacent nucleotide by phosphodiester linkage. There are four key nucleotides and each of it contains deoxyribose sugar, a phosphate group and purine or pyrimidine base. Each nucleic acid contains four types of bases. The two purines adenine and guanine are present in both RNA and DNA. Two types of pyrimidines can be seen in DNA, i.e., cytosine and thiamine. Uracil is replaced by thiamine in RNA molecule.

Two chains of double strand DNA are held together by weak hydrogen bond along the entire length of the molecule. These weak hydrogen bonds primarily provide stability to the DNA molecule. More precisely purine like adenine always pair with pyrimidine thiamine. In a similar fashion the pyrimidine cytosine pairs with guanine. These base pairs are attached to pentose sugar by glycosidic bond. In the two antiparallel strands adenine can base pair with thiamine by two hydrogen bonds and cytosine is paired with guanine by three hydrogen bonds. Presence of three hydrogen bonds between cytosine and guanine makes the helix a more stable molecule. Separation of two strands is more easier at AT than at GC due to its triple hydrogen bond that pairs G and C.

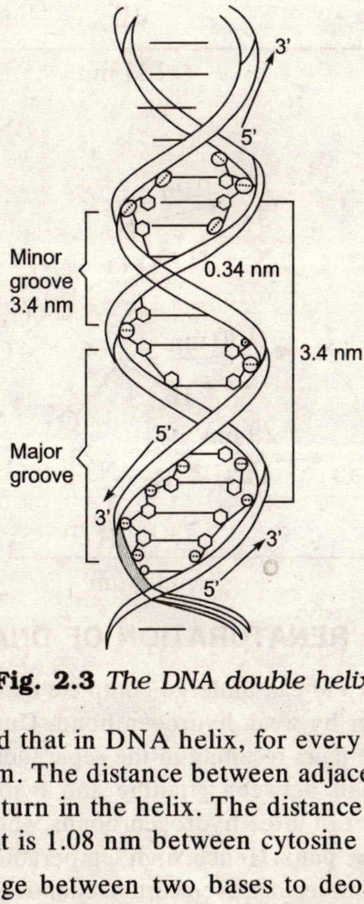

Fig. 2.3 *The DNA double helix.*

X ray diffraction data showed that in DNA helix, for every turns measures 34 Å or 3.4 nm, with a diameter of 20 Å or 2 nm. The distance between adjacent nucleotide is 3.4 Å and there are ten nucleotides present per turn in the helix. The distance between thiamine and adenine base pair in 1.11 nm, whereas it is 1.08 nm between cytosine and guanine.

Presence of glycosidic linkage between two bases to deoxyribose residue is not present directly opposit to each other. As a result sugar phosphate backbone in the double helix of space are not equally spaced along the entire length of the molecule. Grooves formed between the sugar phosphate back-bone is not equal. Therfore major and minor groove is filled with nitrogen and oxygen atom is projected inwards are confined to the top of the base pair. In contrast minor groove can be seen in the double helix. The floor of the major grrove is filled with base pairs, nitrogen and oxygen atoms projected outward to their sugar phosphate backbone.

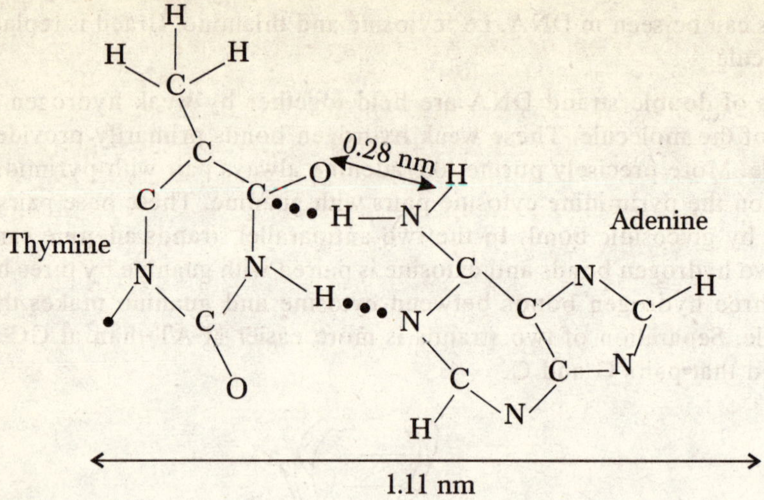

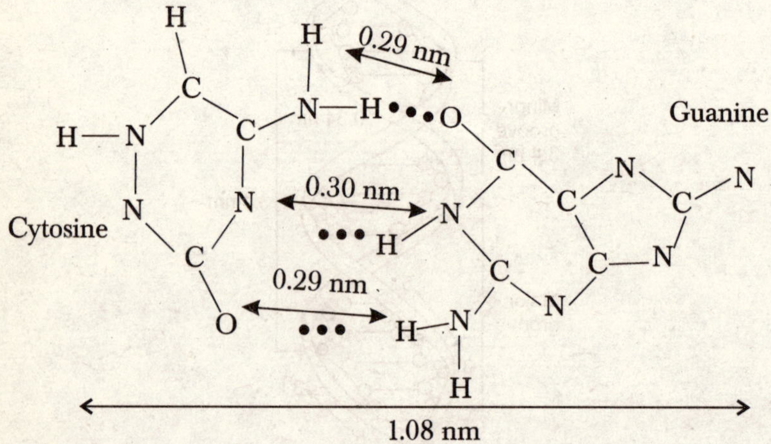

2.6 DENATURATION AND RENATURATION OF DNA

DNA is a highly fragile molecule and can undergo complete denaturation when heated to 100° C. The two strands are held together by weak hydrogen bond, During denaturation hydrogen bond breaks between nitrogenous base pairs resulting in the separation of two strands from each other. During heating process, base pair between guanine and cytosine exert resistance to thermal melting, as they enjoy the support of three hydrogen bonds when compared to double hydrogen bonded adenine and thymine base pairs. Hence, high temperature is indispensable to separate GC rich strand than to break AT rich molecule. Even certain chemicals like formamide or alkali reagent can partially denature DNA. During partial denaturation, strand separation takes place only at AT rich region whereas region rich in GC can retain double helical structure.

Denaturation of DNA is not irreversible. Strand separation due to heating process can be renatured if temperature is slowly reduced. At that temperature single strand often reassociates with their complementary strand and reattains double helical status. In some of the interesting experiments it was shown that when single stranded DNA of man mixed with mice single stranded DNA nearly twenty five percent of these two DNA can form hybrid. It is not suprising that some of the genes of man are similar to the gene sequence of mice. Wherever similar sequence occur hybridization takes place.

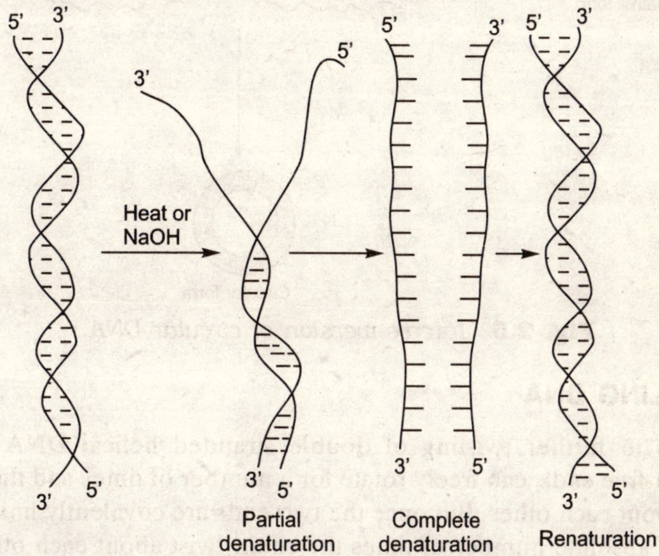

Fig. 2.4 *Denaturation and Renaturation of DNA.*

2.7 CIRCULAR DNA

Evidence based on electron microscopy and autoradiographic technique, several viruses have been studied so far contains circular double-helical DNA molecule. Even short stretch of single stranded DNA in phages are circular. In addition, most of the bacterial DNA are circular. Linear forms of viral DNA can be seen in T_4 and T_7 viruses. Circular nature of DNA pose a problem during their replication. However, this replication problem can be reduced or completely removed by the enzyme which can release the barrier and facilitates replication. It is interesting to see that some of the viral DNA thar are linear outside were found to be circular inside the host cells, proved that these two forms could be inter convertible. Conversion of circular to linear forms is anenzymatic process. A specified enzyme induces two breaks on the circular DNA very close to each other may disrupt hydrogen bond result in transforming circular to linear DNA. Resultant linear DNA contain small stretches of single stranded ends that have complementary sequence or sticky nature at the ends. From these, sticky ends can base pair and resume a circular configuration.

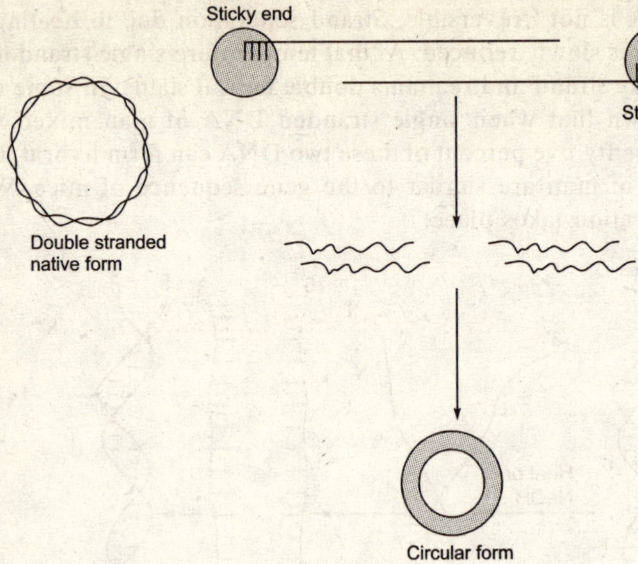

Fig. 2.5 *Interconversion of circular DNA.*

2.8 SUPER COILING DNA

Super coiling refers to further twisting of double stranded helical DNA. The linear DNA molecule due to their free ends, can freely rotate for a number of times and the two strands of the double helix twist about each other. But once the two ends are covalently linked to form circular DNA molecule, the absolute number of times the chain twist about each other cannot change.

Closed circular DNA molecules that are isolated from mitochondria, virus and bacteria are often supercoiled. The configuration of these super coils is right handed. In typical circular DNA isolated from viruses, there is one super coiled twist for every 200 bases. The super coiled structure of DNA is less stable than circular DNA. There is a greater degree of tension mounted on supercoiled DNA. This tension would be reduced if it transforms into circular DNA. The nicking of a single strand instantly converts supercoiled molecule into the simple-relaxed circular state.

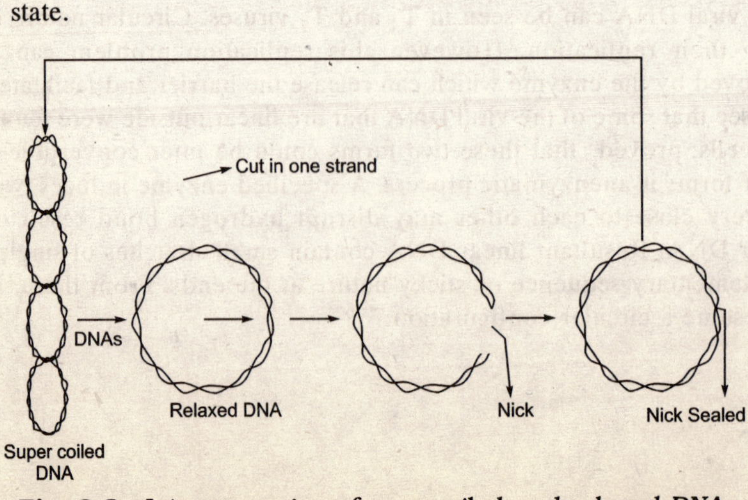

Fig. 2.6 *Interconversion of supercoiled and relaxed DNA.*

Certain parameters are followed to distinguish relaxed DNA from supercoiled DNA by its sedimentation rate. Relaxed DNA moves slowly when subjected to electrophoresis on agarose gel. As position of super coiling is introduced, the DNA becomes increasingly compact.

2.9 SIGNIFICANCE OF TOPOISOMERASES

Topoisomerases are DNA gyrases. This enzyme can convert one form of DNA into another. By changing the linkage number this enzyme converts supercoiled DNA into relaxed DNA and viceversa. There are two types of topoisomerases i.e., I and II. The enzyme topoisomerase II specifically generates negative supercoil. The second type topoisomerase II, which converts the supercoiled DNA into unstrained, energetically more stable relaxed form.

Both the types of topoisomerases (I and II) catalyse the breakage and reconstruction of DNA phosphodiester bonds. These two forms of enzymes create temporary gaps in polynuleotide chains. Topoisomerases are covalently attached to one of the two broken ends in a chain. When a DNA chain is broken by a topoisomerases. They are held together by topoisomerases. In the mechanism of action of topoisomerase I, the enzyme first binds to DNA, undwinding takes place in the double helix. This enzyme nick in one strand and prevent free rotation of the helix by removing bond to each broken end. Later the enzymes passes the other strand through break and ligates the cut end. Thus, increasing the linkage number by one, finally enzyme falls away and the strands renature, leaving the DNA with one negative supercoil.

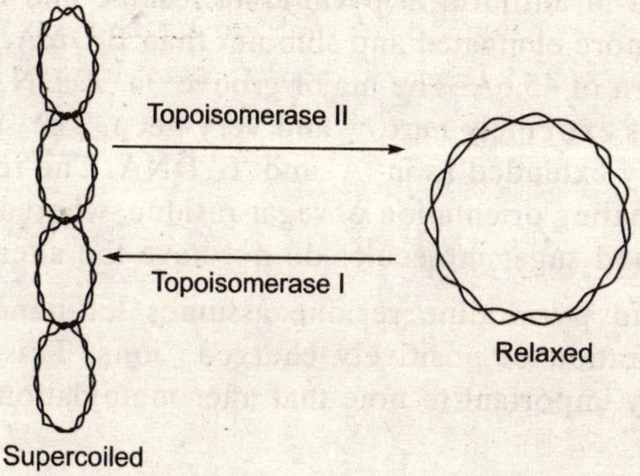

Fig 2.7 *Interconversion of supercoiled DNA by topoisomerase.*

2.10 DIFFERENT FORMS OF DNA

DNA can exits in different forms like A forms, B form and Z forms. The DNA containing alternating pyrimidine residue can turn into purine at left handed as well as right handed helical. There are two forms of right handed DNA, referred to as A form and B form. Most of the DNA in a cell is right handed and exist as 'B' form.

2.11.1 The B-form of DNA

'B' form is the most stable configuration. The 'B' conformation is the longer and thinner forms. 'B' DNA assumes right handed direction. The 'B' form of DNA was proposed by Watson and Crick. It exhibits antiparallel double helix, rotating clockwise i.e., right handed, In solution

DNA usually assumes 'B' form. The 'B' conformation is the longer, thinner forms with 3.4Å thick. Purine and pyrimidine base pairs are oriented perpendicular to the helical axis. 'B' DNA is converted into A DNA by tilting the base pairs to 30°. Therefore base pair occurs at every 2.0Å along the helical axis. In B form the number of base pair per turn helix is 10 whereas base pair per helix repeat is one. The major groove is wide and of intermediate depth. In contrast, minor groove is narrow and of intermediate depth. In 'B' form DNA glycosidic bond is always oriented.

2.11.2 A-form of DNA

The DNA in anticlock position assumes 'A' configuration only when little water is available to interact with the double helix. Relatively unhydrated DNA fibres tend to acquire A form. A-DNA molecules are shorter and broader than 'B' form of DNA. It assumes right handed form. The helix diameter is 25.5Å and can accommodate 11 base pairs per turn of helix. The pitch per turn of helix is 24.6Å. Major groove in 'A' form is extremely narrow but very deep whereas minor groove is very broad but shallow. It is still unclear whether pure A–form DNA exits within cells.

2.11.3 Z-form of DNA

The crystallographic studies have shown that left handed double helices can also exist. This form is called 'Z' DNA. This is due to zig-zag arrangement of phosphate ribose backbone. In 1979, Rich and others obtained Z-DNA in the form of crystals. In 'Z' form left handed DNA in glycosidic bond is oriented in antiform at pyrimidine residue and the 'Syn' configuration at purine residue. Z DNA is more elongated and slimmer than B-DNA. Twelve base pairs can be seen within each helical turn of 45.6Å. The major groove in 'Z' DNA is flattened out on helix surface but minor groove is extremely narrow and very deep. The sugar-phosphate backbone of L.DNA is not completely extended as in 'A' and 'B' DNA. The repeating unit in Z-DNA is a dinucleotide due to alternating orientation of sugar residue, whereas in B-DNA the repeating unit is a mononucleotide and sugar molecules do not have the alternating orientation.

In solutions, purine and pyrimidine residue assumes left-handed configuration in the presence of high concentration of positively charged ions. This nullifies positive charge phosphate group. It is very important to note that after methylation at cytosine base B DNA can convert it into Z form.

In addition several other forms have also been depicted.

2.11.4 C-form of DNA

This DNA could be seen at relatively moderate humidity. It is found at 67 per cent relative humidity in the presence of lithium ions. The number of base pairs per turn in this form is less. It exhibits negative tilt by 7.8°.

2.11.5 D-form DNA

This type is found in extreme cases. The total number of base pairs per turn of helix is eight. They exhibit negative tilt of base pair by 16.7°.

2.12 Viral DNA

The nucleic acids of viruses are either DNA or RNA. Virus consists of double strand (ds) or single strand (ss) DNA. Double stranded DNA replicates by using one of the double strands as template to synthesize partner strand. Viruses with single stranded genome uses the single strand as template to synthesize complementary strand and their complementary strand in turn is used to synthesize its complement which are identical with the original strand. One of the double stranded viruses is SV40 containing six genes of 5,200 base pairs. Vaccinia virus contains less than 300 genes. The total number of base pairs is around 1,87,000. A common T_4 phage contains 165,000 base pairs comprises around 200 genes. Single stranded DNA viruses are also called bacteriophages. The DNA of E. coli, phage are Lambda, $\phi \times 174$ and M 13.

The DNA of $\phi \times 174$, which is one of the smallest DNA virus is circular and contains 5,387 base pairs. The total number of genes are around 11. The DNA of M13 are thin particles which measure 1 μm and their DNA forms an extended loop within the cell. They can infact the cell by adsorbing to an external receptor. This filamentous single stranded DNA phage do not kill the cells or lyse the protein. However, phage particles are extruded continuously as the cell continued to grow.

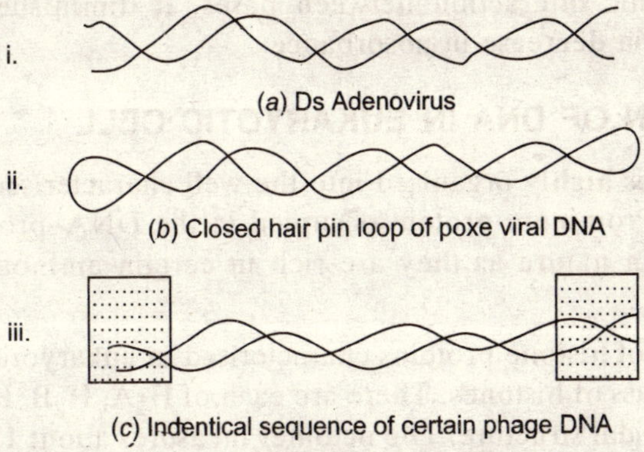

Fig. 2.8

2.13 BACTERIAL DNA

The DNA of bacteria is a huge single molecule packed as one chromosome. It is circular in nature. Generally bacteria contains 2,000–4,000 genes and total number of base pairs is about $< 10^7$. The genes of E. coli and virus are now well studied and better understood than any other forms of life. They provide us all kinds of necessary information on how genetic code operates. DNA of bacteria is large single molecule packed as chromosome. The length of E. coli chromosome is approximately 1.36 nm. Ultra microscopic studies revealed the presence of irregular shaped bodies called as nucleotides. The size of E. coli DNA is about 2.8×10^9 mw and thickness of about 2.0 nm. The total number of base pairs is around 4 million and it is circular form. The bacterial DNA is naked without any nuclear membrane barrier. The DNA molecule assumes super coil status and is around 50 or more highly twisted or suprecoiled loops. The total number of genes on E. coli single chromosome is estimated to be 3000.

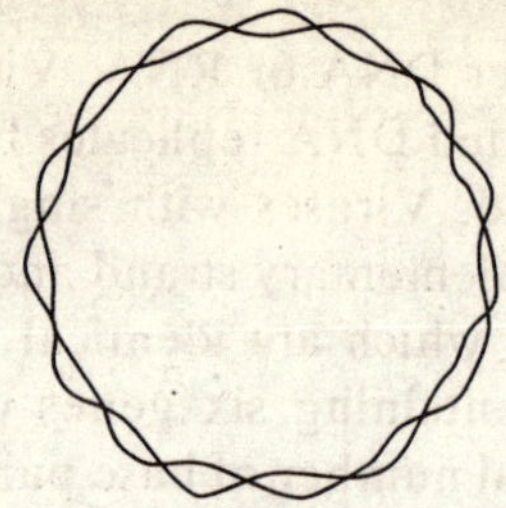

Bacterial DNA

Bacterial contains extra chromosomal DNA called plasmids. Plasmids are small circular double stranded DNA, number ranges between 1–100 copies.

UV absorption of DNA: Nucleic acid absorbs ultraviolet at 260 nm. Absorption potential of DNA depends on the single or double stranded DNA. Single stranded DNA are individual mononucleotides which strongly abosrbs ultraviolet as compared to native double stranded DNA. Increase in absorbency is also due to the number of A-T ratio in the DNA. This phenomenon is known as hypochromatic effect. Less absorption in helical double stranded DNA is due to electronic interaction between bases. It diminishes electronic interaction between bases resulting in decrease in absorbance.

2.14 ORGANISATION OF DNA IN EUKARYOTIC CELL

In eukaryotes the DNA is highly organised into the well characterised DNA protein complex called as nucleosomes. Prominent protein occupied in the DNA–protein complex is histone. These are small, basic in nature as they are rich in certain aminoacids such as lysine and arginine.

There are five classes of histone proteins characterised in eukaryotic cell type. They are H_1, H_2A, H_2B, H_3 and H_4 class of histones. There are each of H_2A, H_2B, H_3 and H_4 (eight histone) forms an octomer ellipsoidal structure. The octomer measures about 11 nm long and 6.5 nm in diameter, This ellipsoidal structure of histone is surrounded by DNA. The DNA coils around ellipsoidal structure of about 165 base pairs. It forms a complex nucleosome structure. Nucleosome is an octomer of four histone protein complexed with protein.

Fig. 2.9 *Nucleosome structure.*

2.15 STRUCTURE OF RIBONUCLEIC ACID (RNA)

Ribonucleic acid exist as unbranched polymeric molecule. RNA molecules could be seen in nucleolus as well as in cytoplasm. The remarkable association between RNA as protein synthesis has been established.

Several refined techniques such as viscocity measurment, light scatering, ultra centrifugation and polyacrylamide gel electrophorsis are commonly employed in the determination of molecular weight of RNA. Most of the plant viruses and few animal viruses contain ribonucleic acid (RNA) as their genetic material.

2.15.1 Primary structure of RNA

Assessment of primary structure of RNA reveals that polymeric molecule is a long single stranded unbranched in nature. Chemically RNA is made up of pentose sugars, phosphate group and nitrogenous bases. The type of pentose sugar in RNA is ribosesugar and it forms the back bone along with phosphate group. The nitrogenous bases are arranged linearly along with phosphate sugar backbone. Therfore chemistry of RNA can be written as nitrogenous bases + Ribose sugar + phosphate = Nucleotide

More importantly, nitrogenous bases or RNA consist of two types, they are purine and pyrimidine. Purine bases are adinine (A) and guanine (a) where as pyrimidine bases comprises cytosine (C) and uracil (U). Replacement of uracil for thiamine is one the diviated characterstics from DNA.

2.15.2 Secondary structure of RNA

In the native RNA, existence of more than one polyribonucleotide strand is ruled out. Therefore, helical region of natural RNA arise from the hair pin like bending of the single strand back upon itself. The interacrion between A and C and G stabilizes helical region. It the proposed postulates on it non-helcal and non-helical organisation of RNA, nucleotide may occur largely as numerous small loops in absence of its participation in the helical organisation where as non-helical RNA take the form of extended amorphous zone separating segments of helical organization.

Ribozyme

According to the central dogma of biology, nucleic acids (RNA × DNA) are heredity molecules, where as proteins are the molecules of function and biocatalysis. However, in 1967, it was suggested that RNA could exhibit catalytic property. This cataltic RNA is popularly known as ribozyme. In 1978, Altman studied this RNA cutting enzyme from the bacterium and named as RNAsp consist of a complex between single protein and single RNA molecule. In an extended work RNAsp was splitted to separate protein from the RNA. As a result the enzyme was no longer functional. However, researchers restored the enzymatic activity by removing various components. This evidence was proved that first time an RNA molecule had been to shown to be indispensable for a catalytic reactions.

In the real sense, ribozymes are antisence RNA molecules with catalytic activity. The enzymes whose RNA subunits carry out catlatytic reactions are called ribozymes. The most important property of the ribozyme specificity to cleave a unique target. They may catalyse intra molecular (Seit-cleavege) clevage as well as intermolecular cleavage (external cleavage). In ribozyme cutting reactions inter and intra molecular cleavage involvues joining or cleavage of phosphodiest bond.

Hammer head Ribozyme

The striking similarity between the shape of hammer head and its secondary structure gained this name. The hammer head ribozyme is an antisense RNA with a small catalytic RNA motif. It is involved in the replication of a type of viroid and some satellite RNA the secondary structure of ribozyme includes stem and loop feateves where as structure resembles a wish bone, stabilized by Watson Crick base pairs.

The catalytic property of hammer head ribozymc is dependent on pH. The ribozyme exhibit fully active at pH 8.5 when compared to the pH. 5, where it is totally non-functonal or show. Hammer head ribozyme motif is known to be involved in viroid replication by catalysing the cleavage of longer position strand of viroid genome.

Catalytic mechanism by ribozyme :

The hammer head ribozyme catalysis a transesterification reaction in which the 3′, 5′-phosphodiesterbond between nucleotides 17 and 1% is cut. This catalysis mode of ribozyme is called as in-line mechanism. It requires Mg2+ cation to carry out this process.

Hairpin Ribozyme

These are small catalytic RNAs cleaves RNA substrates. The negative strand of tobacco ring spot virus satellite RNA was found to contain hairpin catalytic motif. Which mediates self-cleavage and ligation reactions. The hairpin ribozyme acts as better better ligase known as nuclease. Where as the hammer head ribozyme reaction involves cleavage rather than ligation process. The satellite RNA of plant virus contains hairpin ribozyme up to 50 nucleotide long and cleave itself internally can also cleave other RNA strand in a transtesification reaction.

RNAase p:

This was the first demonstrated ribozyme RNA with catalytic activity dut to the complete dissection of ribonuclease p. RNAase p can be dissociated into its two components. The 375 base RNA and the 20KD polypeptides. Invitro experiments proved that both components were necessary to cleave the RNA substrates in this RNA protein complex, RNA alone can catalyse the reaction by cleaving multiple substrates. However, protein component in association, can greatly increases the speed of the reaction. Therefore confirming the role of both the component for natural enzyme activity.

Tharapeutic Significance of Ribozyme

Ribozyme technology has been used in reducing the levels of specific intracellular proteins offers multiple therapeutic opportunities. This strategy is known to be used in some therapy. One of the most attractive therapy using ribozyme technology is in the treatment of HBv (Hepatitis B virus) infection.

2.16 STRUCTURE AND TYPES OF RNA

Transcription of protein coding and RNA coding genes (rRNA and tRNA) are facilitated by RNA polymerases. Transcription process generates different types of RNA like messenger RNA (mRNA) transport RNA (tRNA) and ribosomal RNA (rRNA).

2.16.1 Messenger RNA (mRNA)

Synthesis of mRNA is due to the transcription of protein coding genes. Transcription and translating are the two indispensable process closely linked and occur simultaneously. The ribosomes can attach to mRNA even before the completion of transcription.

Prokaryotic mRNA are more stable than eukaryotic mRNA. Due to the lack of compartmentalization translation of mRNA takes place within the visinity of bacterial chromosomes. In contrast compartmentalisation of nucleus and cytoplasmic space in the eukaryotic cell would force the mRNA to undergo translation only in cytoplasmic region. In E. coli cells there are 600 differenct types of mRNA, all together constitute 1,500 moles per cell, at least 2–3 copies were found for each type. mRNA represents only a small percentage (2–4%) of the total cellular RNA. In bacteria the half life of mRNA molecule is about 1–2 minutes. Since mRNA is known for its instability can undergo degradation in 5′–3′ direction i.e., in the same direction as the synthesis of mRNA and their translation. In bacteria, rapid synthesis and degradation helps in adaptation. Eukaryotic mRNA exhibit considerable stability and they can exist for longer time. The half-life of mRNA of mammalian cell in culture ranges from few hours to 24 hrs. Availability of mRNA for protein synthesis may be from few hours to days. The prokaryotic mRNA is synthesised as long poly-cistronic mRNA i.e. many strucural genes are transcribed into mRNA as a single unit. In equkaryotes its transcription product as monocistronic mRNA i.e. single gene mRNA. A poly-cistronic mRNA as a result of uninterrupted transcription of about ten adjacent genes in an operon. Thus it can be 3 times longer than the ribosomes and 120 times longer than the tRNA. All mRNA contain two specific region begins with initiation codon AUG and the termination condon UAG, UAA nad UGA. It also contain leader region in upstream of initiation codon and downstream direction i.e. 3′ and other is tail region.

2.16.2 Transfer RNA (tRNA)

Transfer RNA (tRNA) is an adapter molecule occupies nearly 10–15 per cent of the total RNA in the cells of E. coli. This RNA is a polynucleotide contains between 75–95 nucleotides in length and play a important role in protein synthesis as an adapter molecule. In E. coli there are more than 50 different tRNA able to utilise 61 codons among possible 64 triplet codons of genetic code. Some tRNA can able to recognise more than one codon. The tRNA can bind covalently to their corresponding amino acid through a sequence, CCA at 3′ end. This entire complex is known as aminoacyl-tRNA and it facilitate particular aminoacid to brought at the ribosomal site. The proposed clover leaf model of tRNA exhibit following features.

(a) Tranfer RNA contains four arms, they are acceptor arm TψC arm, anticodon arm and the 'D' arm.

(b) The accepter arm contains CCA at 3′ end which are unpaire. This can be attached to amino acid. This attachment site is known as acceptor end.

(c) TψC arm contains triplet sequence in which modified base known as pseudouridine is present (ψ symbol stand for pseudouridine).

(d) The anticodon arm or loop is present exact opposite from the acceptor end or arm. The anticodon form triplet bonds with mRNA codon.

(e) The 'D' arm, which contains dihydrouridine.

(f) All tRNA exhibit secondary and tertiary structure and share certain common feature.

(g) tRNA assumes a common clover leaf secondary struture according to proposed model.

(h) Most important variable feature of tRNA is a presence of extra arm based on this two classes of tRNA. The class of 1 rRNA has a small arm contains only 3–5 bases. Class 2 tRNA has large extra arm consist of 13–21 bases.

(i) tRNA is linked to aminoacid by an ester bond. The secondary structure of tRNA folds into L shape.

(j) The tertiary structure in which the side arms are folded together with the anticodon at one end of the L and acceptor end at the other.

(k) The 'D' and T loops are held by hydrogen bond. All tRNA show common feature of these characteristic folding. However, mitochondrial tRNΛ exhibit sometime slight different characteristic folding.

(l) The accuracy of protein (Attachment between tRNA and aminoacid).

(m) The bond formed between tRNA and aminoacid and CCA end is highly energised by storing energy from the ATP which could be utilised in the later stage to form peptide bond at ribosome site (Fig. 2.1)

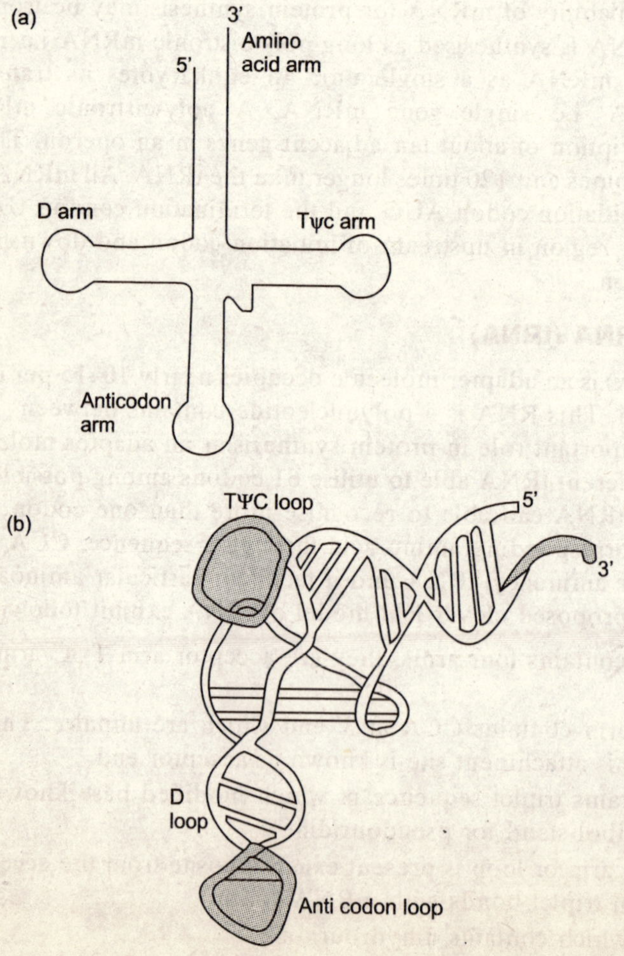

Fig. 2.1 *Secondary (a) and tertiary (b) structure of tRNA.*

(n) tRNA genes transcribed by RNA polymerase III in eukaryotic cells. tRNA genes have a two separate promoter region known as A and B boxes. Atleast two transcription factors must bind to these sequence before tRNA genes are recognised by RNA pol III.

(o) After transcription of tRNA it undergo some post transcriptional modification. For Ex. in E. coli tyrosine tRNA containing 3 extra nucleotide after transcription. These extra nucleotide are removed to obtain active tRNA molecules. Other tRNA are synthesised as long precursor that contain 5 to 7 RNA in a single RNA chain. Therefore these nucleotides are cleaved by ribonuclease enzyme during processing.

2.16.3 Ribosomes

It is a granule like ribonucleoprotein particle consisting about 65 per cent RNA and 35 per cent protein. These are found in all the cells involved in protein synthesis. A tiny bacterial cell contains 10,000 ribosomes which represents nearly 25 per cent of the total mass of the E. coli cells. The molecular weight of ribosomes is about 3 million. In mammalian cells the total number ribosomes may reach 10 millions. The ribosomes of eukaryotes are generally larger than its prokaryotic counterpart (double size).

Ribosomes have two subunits one large and other is small subunit. The ribosomes of prokaryotes (bacteria) are smaller in size and are called 70S due to their sedimentation constant. They have subunit of 30S and 50S when mg^+ ion concentration in the solution is 0.01 M. The 70S subunits of prokaryotes can dissociate into 50S and 30S particle when mg^{2+} concentration is lowered to 0.001 M. The ribosomes of eukaryotes are slightly heavier called 8S alone have two subunit of 40S and 60S. They can dissociate into 60S and 40S particles, when Mg^{2+} ion concentration is lowered. Ribosomes are also found in cell organelle like mitochondria and chloroplast of eukaryotes and are actually 70S status. They are smaller than eukaryotic ribosomes. The size of organelle ribosome can be comparable to the prokaryotic 70S ribosomes.

2.16.4 Structure of Ribosomes

The prokaryotic ribosmoes are called 70S comprises of smaller 30S and larger 50S subunit associated to form the 70S ribosome. According to Stoffar and Willmann model the long axis of 30S subunit is oriented to the central protruberance of 50S subunit. A tunnel is formed between smaller and the larger subunit. The mRNA binds to the ribosomes and passes between right temporal region of 30S subunit contains 16S rRNA of molecular weight 550,000 and nearly 21 proteins. However, 50S subunit contains 23S and 5S ribosomal RNA with molecular weight 1,100,000 and 40,000 respectively and nearly 34 proteins (Fig. 10.2).

2.16.5 80S ribosome

Eukaryotic ribosomes are called 80S comprises of smaller 40S and larger 50S subunit. They are associated with large number of proteins and molecules of RNA. The 40S subunit is curved and ellipsoidal consist of head and a body. It contians molecules of 18S RNA and 30 different proteins (Fig. 10.3). It constitute 45 per cent of the subunit. The larger 60S subsunit contains 28S, 5.8 and 5S ribosomal RNA and 40 proteins of molecular weight between 11,500 to 41,000. The RNA constitute nearly 59 per cent of the subunit.

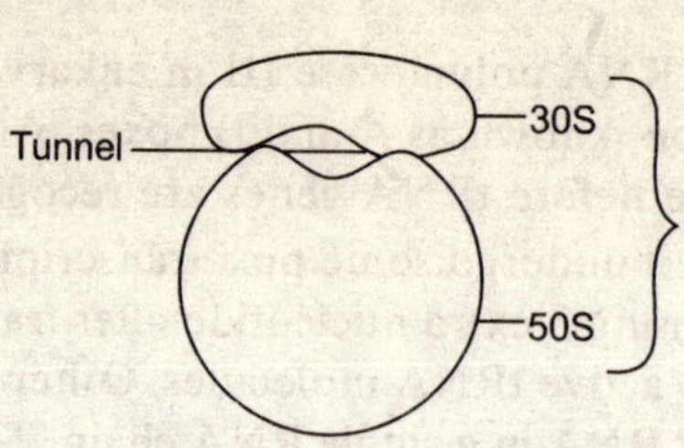

Fig. 2.11 *Prokaryotic Ribosome.*

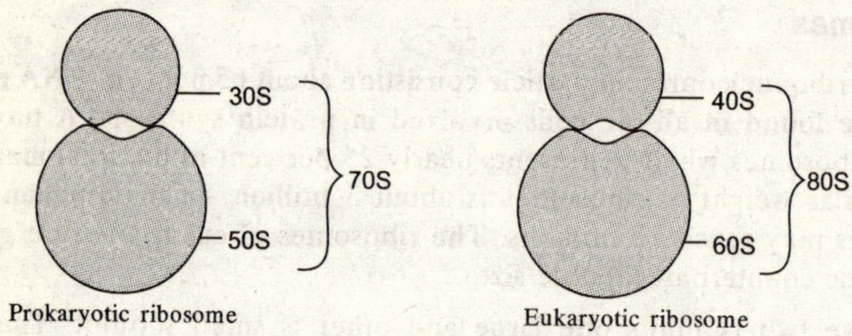

Fig. 2.12 *Ribosomes Structure.*

2.16.6 Polysomes

Polysomes are group of ribosomes. These are also called as polyribosomes. Protein synthesis does not takes place on a single ribosomes but on polyribosomes. They may be arranged in a linear fashion or rosette shaped cluster. The number depends on the length of the mRNA molecules.

2.16.7 Functions of Ribosomes

(a) Take part in protein synthesis.

(b) The ribosome functions as a template, bringing together different component of protein synthesis.

(c) 16S rRNA provides a binding site for tRNA.

(d) Ribosome have a protective function. It prevents mRNA degradation from the enzyme nucleases.

SUMMARY

DNA is blue print of life. It is present in all linving cells. Unravelling DNA and its detailed structure have been well established by the contribution of many scientists. Much painstaking work through X-ray diffraction was done by Watson and Crick. They proposed a model on double stranded helix. Some scientists like. Avery and Taylor signifies that DNA is a genetic material based on established experiments. Watson and Crick's double helical model exhibit B form of DNA and they replicate in semi-conservative model. Chemically DNA consists of four nitrogenous bases, pentose sugar and phosphate. The nitrogenous bases are organised in purine and pyrimidine based on their ring structure, pentose sugar in DNA consist of five carbon atom of which four carbon atoms combine with one oxygen atom and forms a ring.

Phosphate group present in the DNA contribute strong negative charge. Purine group contains Adenine (A) and Guanine (G) whereas pyrimidine comprises of Thiamine (T) and Cytosine (C). During helical formation Adenine always pairs with thiamine (A = T) by the formation of two hydrogen bonds and Guanine pairs with cytosine with the help of three hydrogen bonds (G = C). Thus, purine bases pair with pyrimidine bases. In DNA the amount of adenine equals amount of thiamine. This is known as Chargaff rule. During organisation of DNA, nitrogenous bases and sugar makes complex known as nucleosides and this nucleosides joins with phosphate to form nucleotides. These nucleotides are interlinked by phosphodiester bond and in the process of polymerisation become polynucleotides.

The two chains are held together by weak hydrogen bonds. However, hydrogen bond is contributes by maintaining the stability of the DNA molecule.

DNA can undergo denaturation and renaturation as it is highly a fragile molecule. It can undergo complete denaturation when heated to 100° C. Denaturation is due to breakage of hydrogen bonds. However, base pair between guanine and cytosine exert resistance to thermal melting due to hydrogen bond. Nature of DNA in bacteria and viruses are circular and double stranded. DNA exhibit super coiled status. There are different forms of DNA seen in the cell. Thses exist as A form, B form, Z form and C form.

There are three types of RNA take part in protein synthesis. There are 600 molecules of mRNA can be found in bacterial cell. mRNA contain initaiator codon and termination codon. Transport RNA contain four arms as exhibit secondary plus tertiary structure. tRNA genes are transcribed by RNA polymerade III. Riboforms are tiny cell organella take part in protein synthesis. It is made up of RNA and proteins. On prokaryotes ribosomes are 70S comprites of 30S and 50S subunits. Eukaryotes contains 40S and 60S together constitute 80S subunits ribosome.

QUESTIONS

1. Explain how the important features of Watson and Crick model explained the known properties of the DNA molecule. (BU, May 1996)
2. Explain the structure of DNA and its functions. (BU, April 1998)
3. Explain the organisation of double redical DNA.
4. Describe different forms of DNA.
5. Explain the structure and role of tRNA. (BU, April 1998)
6. What are different forms of RNA ? Give function each. (BU, Nov. 1999)
7. Describe the structure and functions of ribosome. (BU, April 2000)

SHORT QUESTIONS

Explain the follwing terms : (BU, 2000)

1. z-DNA
2. Forms of DNA
3. Super coiled DNA
4. Avery experiment
5. Structure of Ribosomes
6. mRNA.

3

CHAPTER

DNA Replication

3. INTRODUCTION

DNA replication is one of the most fascinating and fundamental properties of all living cells. It can be seen that replication of DNA can contribute to the biosynthesis of DNA. The process of duplication can be observed at several levels. The genetic material is duplicated by replication during cellular division. Replication of DNA is indispensable for the cells. This capacity for self duplication paves a critical way for the evolution of primitive life on this planet. Although DNA molecule contain genetic information for its replication, it lacks the ability to do so by itself. Several essential factors actively participate in order to achieve DNA replication. An impressive study has shown that several proteins and key enzymes participate and are committed to perform DNA replication at the backdrop of evolution of biomolecules. RNA was the first genetic material to undergo simple replication process. However, during the course of evolution, RNA as a genetic material gave away its responsible status to DNA and further complicated the replication process due to involvement of several auxiliary material.

Replication involves major disruption of the structure of DNA. Although two strands of the DNA must separate, they do not exist as a single strand. Therefore separation is only a temporary process and is reversed as the daughter duplex is formed. The double helical structure proposed by Watson and Crick is well suited and clearly explains the mechanism of replication of DNA. The two strands of the double helix can separate because they are held together by relatively weak hydrogen bonds. Each polynucleotide chain acts as a template for the synthesis of a new DNA molecule. Once the replication is complete base pairing rules imply between one of the parental strand and another from newly formed daughter strand of DNA. During base pairing Adenine (A) pairs only with Thiamine (T), and Guanine (G) with cytosine (C), and as a consequence each daughter molecule is an exact replica of the parental molecule.

3.1. DNA REPLICATION IS SEMI CONSERVATIVE

It was suggested that DNA replication is semi-conservative. In the conservative mechanism one of the strands of parental DNA molecule serves as a template for the synthesis of the complementary strand. More precisely 'semi-conservative' means half of the DNA is conserved i.e., only one strand is synthesized and the other half of the original DNA is retained. The specificity of base pairing clearly suggests that each of the separated parental strand could act as template for the synthesis of a complementary daughter strand. As a principle the newly formed daughter strand is assembled on each parental strand. During semi conservative replication the sequence of daughter strand is totally controlled or dictated by parental strand, for example, 'A' in the parental strand places 'T' in the daughter strand, similarly 'G' on the parental strand directs the fixing of 'C' on the daughter strand. The term semi-conservative was named because each strand on the original (parental) molecule serves as a template for synthesis of complementary strand. As a result, after duplication cycle, there are two molecules of hybrid DNA strand in which one strand is parental and another strand is the daughter molecule (Fig. 3.1).

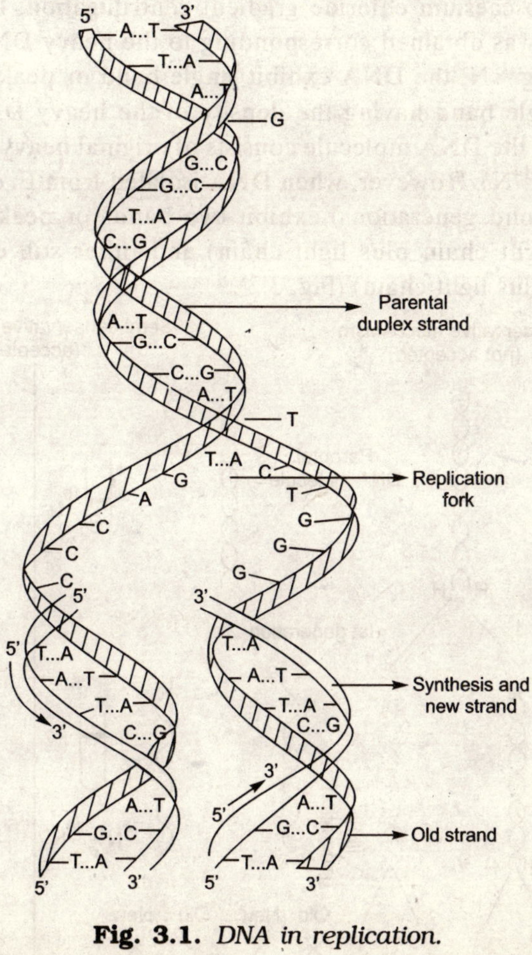

Fig. 3.1. *DNA in replication.*

In the conservative model approach the two original strands would remain together or conserved and two daughter structures are newly synthesized. In another approach, replication is known as dispersive model, the over all integrity of each of the parental strand becomes disrupted. According to this model, as a result of disruption of parental strand, the daughter cell contain duplexes in which out of the strand is old and another is new DNA and this model claims that neither the strand nor the duplex itself is conservative.

3.1.1 Experimental Evidence

In the absence of information responsible for replication additional experiments have been documented. Following are some of the key experiments conducted to evidence that DNA replication is semi conservative.

3.1.1.1 *Meselson and Stahl Experiments*

Experiments conducted by Meselson-Stahl shows that the DNA replication takes place by semi-conservative mechanism. In this classical experiment they employed radioisotopes ^{15}N and ^{14}N in *Escherichia coli* as a model system. First *E. coli* was grown in medium containing ^{15}N heavy isotopes ($^{15}NII_4Cl$ as a sole source of nitrogen) for several generations. It was then shifted to another medium containing ^{14}N isotopes ($^{14}NH_4Cl$). At regular intervals DNA is then isolated and subjected to caesium chloride gradient centrifugation. It was found that at time zero only a single band was obtained corresponding to the heavy DNA. After one generation in the mixture containing ^{14}N, the DNA exhibit single band or peak in the absorbency. This clearly suggests that single band having the density of the heavy DNA and that of the light DNA. Which means that the DNA molecule consists of original heavy strand (^{15}N) and a newly synthesized light strand (^{14}N). However, when DNA isolated from *E. coli* grown in the medium for two generations (second generations) exhibit two bands or peaks of DNA in which two strand contains ^{14}N (Light chain plus light chain) and other still corresponding to hybrid molecule (heavy chain plus light chain) (Fig. 3.2).

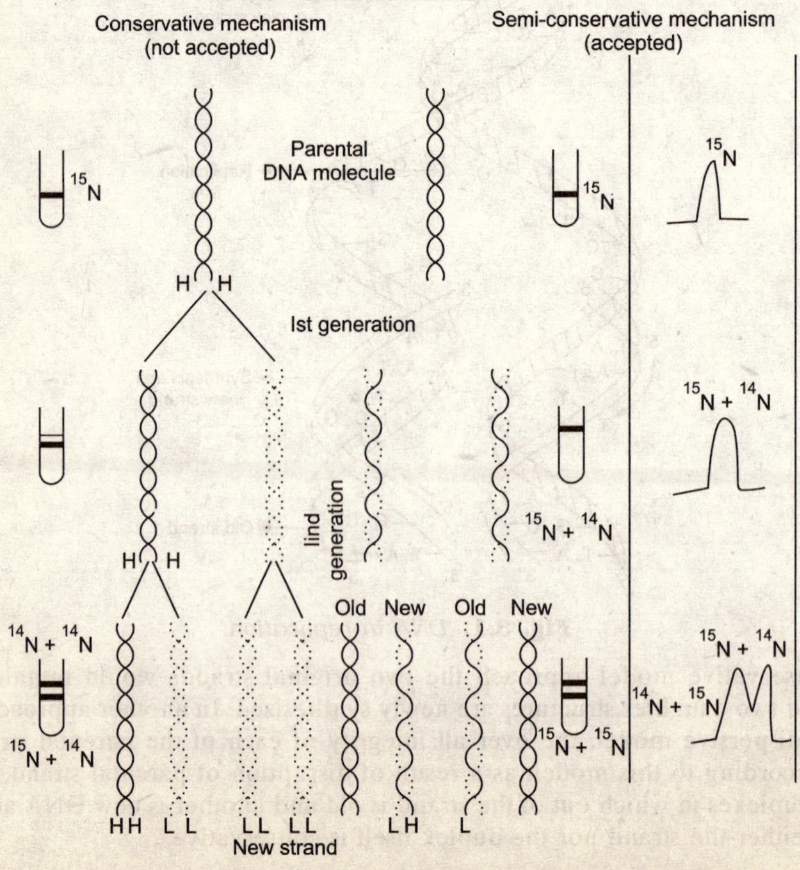

Fig. 3.2. *Semi-conservative replication of DNA.*

3.1.1.2. *Cairn's Model*

J. Cairn's (1963) employed autoradiography technique to visualize replicating chromosome of E. coli. In the autoradiography technique, sample material is fixed with radioactive material like tritiated thymidine. It is a heavy isotope of hydrogen which replaces normal hydrogen in the thymidine to give rise to tritiated thymidine. The advantage of triated is that it will selectively label only DNA but not RNA due to its thymine absence. The autoradiography was done at regular intervals and confirmed that DNA replication is in semi-conservative mode. This study confirmed that the replication begins at a specific site called origin and moves in one direction within a replication fork. In circular DNA, denaturation takes place at a specific point. It is followed by bi-directional synthesis of DNA. This bidirectional synthesis mode travels in opposite direction around the circular DNA. Unwinding of DNA follows continuously along with growing points. The unwinding of DNA creates a torque that is transmitted to the unreplicated part of the DNA molecule resulting in the formation of supercoiled status. Supercoiled status of DNA prevents further replication. The problem is solved by temporary nick on one strand of the DNA by swivelling protein. This allows the parental strand to undergo free rotation on each other and finally get freed. As the replication process proceeds the two growing points converge on the terminus (Fig. 3.3).

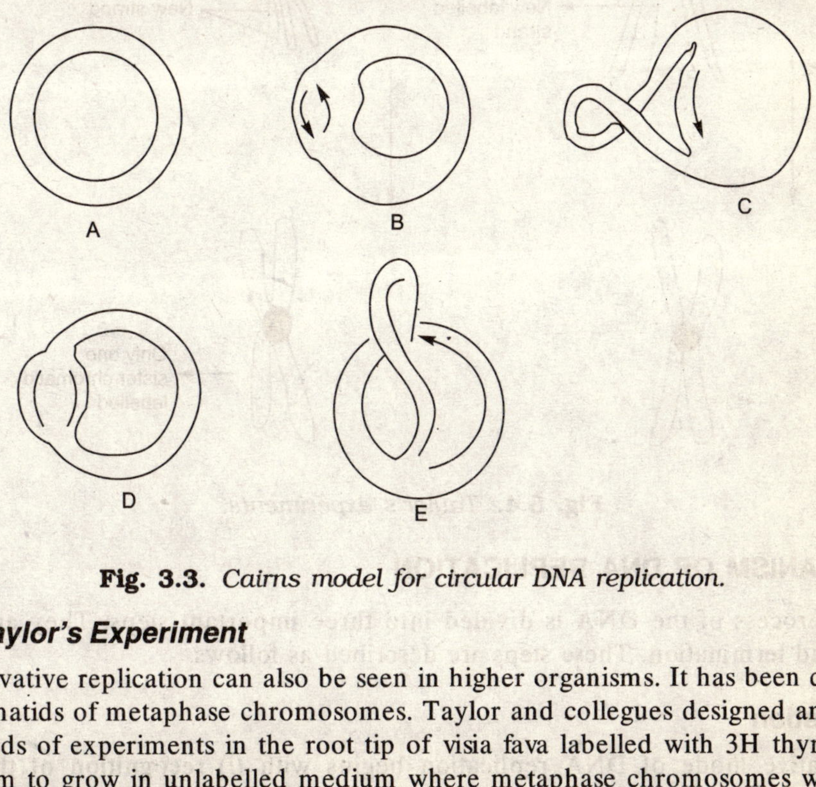

Fig. 3.3. *Cairns model for circular DNA replication.*

3.1.1.3 *Taylor's Experiment*

Semi-conservative replication can also be seen in higher organisms. It has been demonstrated in the chromatids of metaphase chromosomes. Taylor and collegues designed and conducted different kinds of experiments in the root tip of visia fava labelled with 3H thymidine. They allowed them to grow in unlabelled medium where metaphase chromosomes were analysed by autoradiography. It was found that both the sister chromatids were labelled after one generation (first division). However, when the cell undergoes second division in unlabelled medium, only one sister chromatid was labelled. This clearly evidenced semi-conservative DNA synthesis (Fig. 3.4).

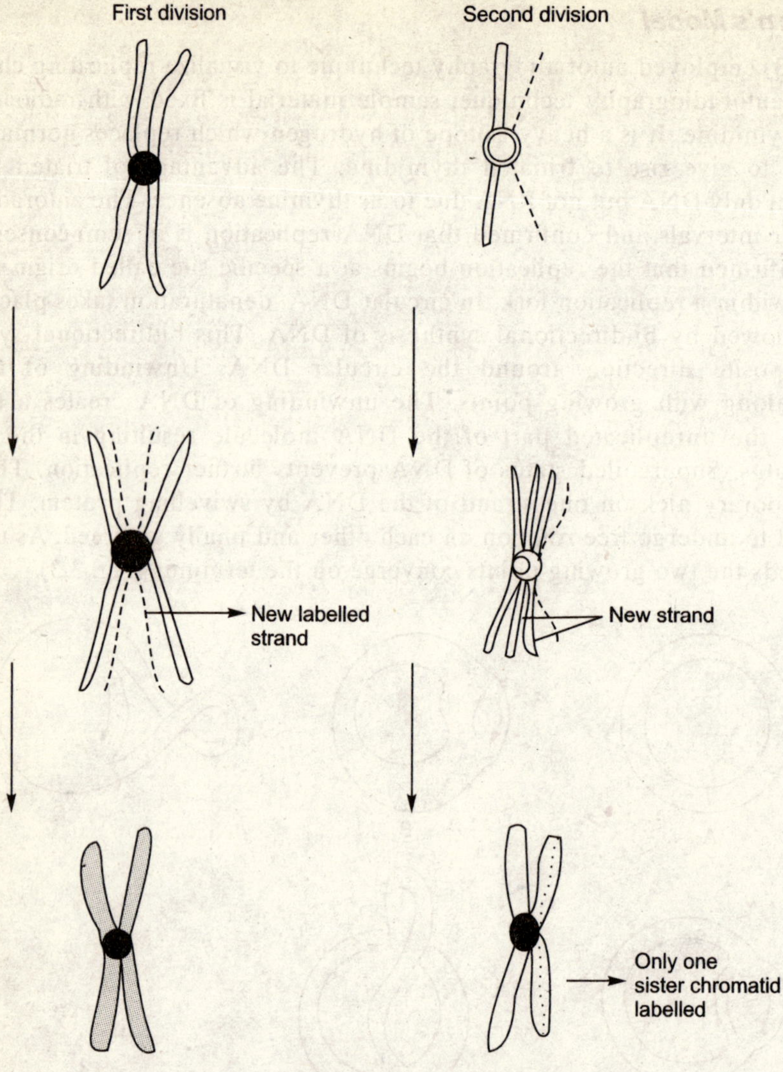

Fig. 3.4. *Taylor's experiments.*

3.2 MECHANISM OF DNA REPLICATION

Replication process of the DNA is divided into three important steps. They are initiation, elongation and termination. These steps are desoribed as follows.

3.2.1. Initiation

Semiconservative mode of DNA replication begins with (i) recognition of the origin of replication point called as replicans (ii) which is followed by strand separation (iii) synthesis of RNA primers (iv) synthesis of initiation of small DNA fragments on the RNA primers, leading to replication fork.

Replication begins at a specific site on the bacterial chromosome called the origin of replication (Fig. 4.6). Detailed studies on the E. coli DNA replication reveals that there is only one origin of replication known as ori c. The recognition of origin of replication is done by proteins. Some specific proteins prevent initiation of replication at sites other than this ori c. The first step in the initiation of replication is the specific binding of ` protein to ori c. The protein dna coded by DNA is complexed with ATP, will open the double helix. The separation of two strands by another protein coded by a gene dna which introduce a strain on the DNA resulting in few more twists on the whole DNA. This is known as supercoiling. These supercoils can be relieved into relaxed form by an enzyme DNA gyrase. This enzyme belongs to the topoisomerase capable of converting DNA helix into supercoils and vice versa. In addition, specific proteins known as protein HV could prevent initiation, replication at sites other than ori c by eliminating RNA primers synthesized at illegal sites. All these proteins are committed to facilitate replication and are together known as replisomes. As the replication initiates, hydrogen bonds between bases of two strands get broken down resulting in continuous unwinding of double helical DNA. Unwinding of DNA is done by DNA helicases. Once replication starts, replication proceeds bidirectionally along the entire length of DNA (Fig. 3.5).

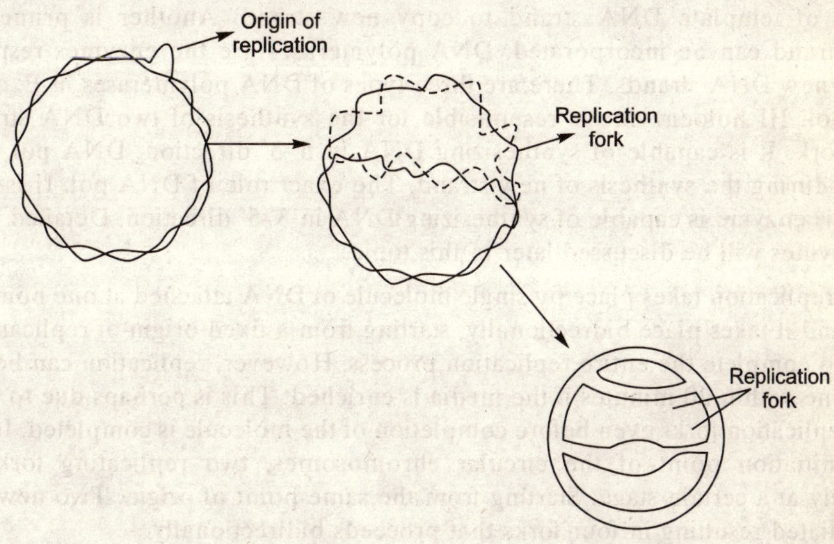

Fig. 3.5. *Bidirectional replication in circular chromosome.*

3.3 ELONGATION

3.1.1. Replication Forks

Replication fork is a point at which pair of replicated segments come together and join the non replicated segments. Replication fork can be seen at the site where the parental double helix is undergoing strand separation and nucleotides are being inserted to the newly synthesized complementary strand. Replication proceeds outwardly from the origin in both the directions i.e., bidirectionally. The two replication forks moves in opposite direction proceeds and finally meet at the point of origin where replication terminates. The replication fork moves relatively at a faster rate i.e., 500 to 1000 nucleotides per second.

When DNA strands are separated into single stranded zones, these single stranded DNA must be prevented from reassociation of strands into duplex. These functions are carried out by a special protein called single stranded DNA Binding protein (SSB). DNA multiplication always takes place in 5′-3′ direction in both the strands of DNA, semi-conservative replication proceeds by the synthesis of single stranded DNA alongwith replication fork. DNA polymerase cannot initiate DNA synthesis by itself, but it can add deoxyribonucleotide on RNA primer and proceeds with synthesis of DNA.

3.3.1.1 *RNA Primers*

These are small strech of RNA fragments generated by DNA-directed RNA polymerase known as 'primases'. The first step in the replication of DNA is the synthesis of RNA primer. The small primer strand is complementary to DNA and 50-100 nucleotides long. Once primer is synthesized by primases, the DNA polymerases begin to add deoxyribonucleotide to 3¢-terminal of the RNA primer.

All the DNA polymerase involved in replication have two basic requirements. One is the requirement of template DNA strand to copy new strand. Another is primer to which nucleotide strand can be incorporated. DNA polymerases are the enzymes responsible for constructing new DNA strands. There are three types of DNA polymerases in E. coli bacteria. The DNA pol. III holoenzyme is responsible for the synthesis of two DNA strands at the replication fork. It is capable of synthesizing DNA in 5′-3′ direction. DNA pol. I facilitates DNA repair during the synthesis of new strand. The exact role of DNA pol. II is not known. However, this enzyme is capable of synthesizing DNA in 3′-5′ direction. Detailed information on these enzymes will be discussed later in this topic.

In E. coli, replication takes place by single molecule of DNA attached at one point to the cell membrane and it takes place bidrectionally, starting from a fixed origin of replication. It takes 30 minutes to complete the entire replication process. However, replication can be completed at a lesser time within 20 minutes if the media is enriched. This is perhaps due to the onset of many new replication forks even before completion of the molecule is completed. It was shown that from initiation point of the circular chromosomes, two replicating forks proceeds bidirectionally at a certain stage. Starting from the same point of origin. Two new replication forks are initiated resulting in four forks that proceeds bidirectionally.

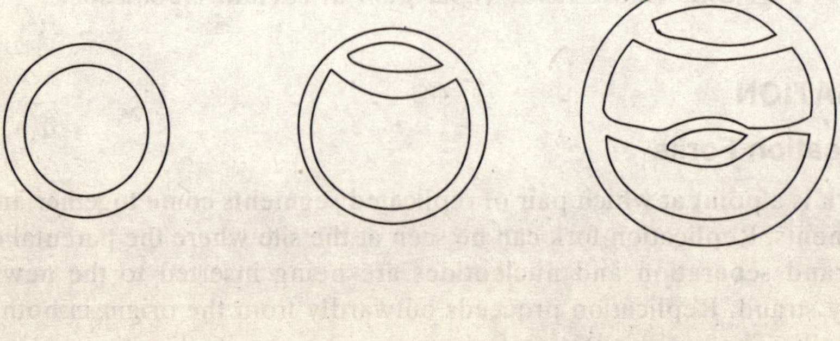

3.4 SEMI DISCONTINUOUS REPLICATION

It has been shown that replication is continuous on one strand and discontinuous on the other. DNA strand cannot be synthesized in 3′–5′ direction rather both newly synthesized DNA strands proceeds in 3′–5′ direction rather both the newly synthesized DNA strands proceed in 5′–3′ directions. During progression of replication fork, one of the newly synthesized strands grows towards the replication fork where DNA strands are being separated. While the other strand synthesis takes place away from the replication fork or against replication fork. It is quite simple that one of the DNA strand can be synthesized by DNA pol III in 5′–3′ direction using 3′–5′ DNA as template. However, DNA pol. III cannot synthesise DNA strand in anti parallel 5′–3′ strand. Then it is amazing to see how the other strand is synthesized. Several consolidated evidences have shown that the strand that synthesizes away from the replication fork is synthesized discontinuously like fragments. Hence the strand that is synthesized continuously is called the leading strand. In the contrast the strand that is synthesized discontinuously in opposite strand is called lagging strand. Thus DNA replication is semi discontinuous. These small fragments were discovered by Okazaki, Nagoya University, Japan using various labelling experiments. These small DNA fragments are known as "okazaki fragments" after his name (Fig. 3.6).

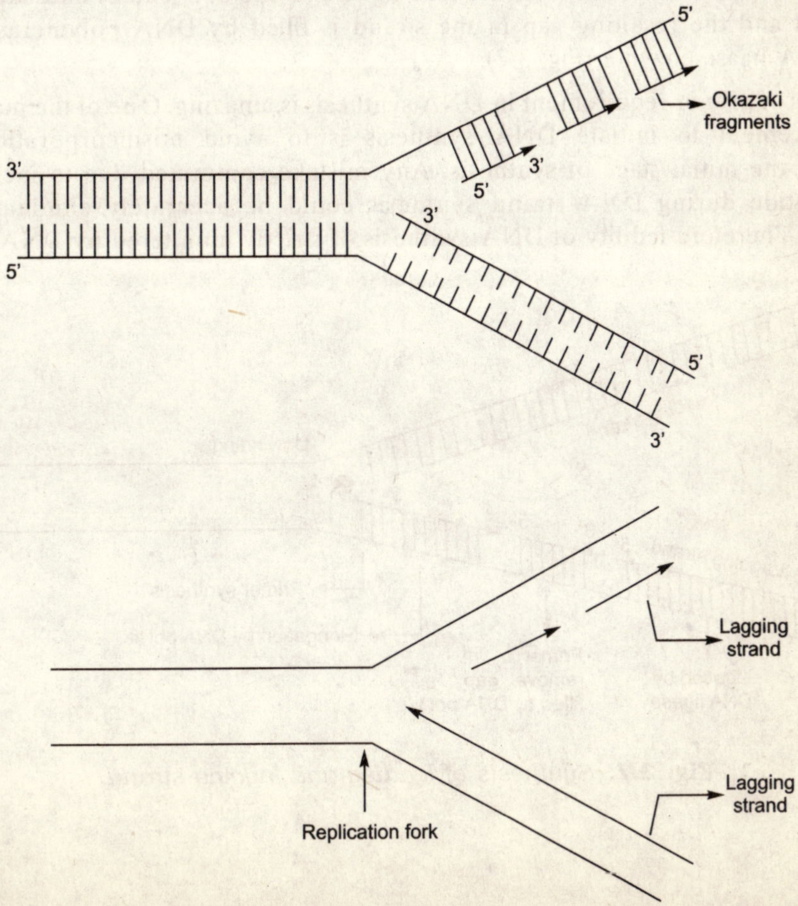

Fig. 3.6. *Shows replication fork and the formation of okazaki fragments.*

Initiation, elongation and completion of okazaki fragments occurs very rapidly. The recycling of the polymerase to the next okazaki fragments would require that the enzyme move towards replication fork. After careful studies it was thought that the enzyme does this by hitching ride with replication machinery on the leading strand template. Thus, although two polymerases are moving in opposite direction along with their respective strands, they both act as they were part of the same protein complex.

3.5 ENIGMA OF DNA REPLICATION

For several years initiation and formation of each of the DNA fragments was an enigma, since none of the DNA polymerase is capable of strand initiation. It was soon discovered that initiation of DNA strand is not accomplished by DNA polymerase but rather by an enzyme known as primase that constructs a short primer composed of RNA.

The leading strand, whose synthesis begins at the origin of replication is invariably initiated by RNA polymerase. Once this primer is synthesized which is then extended by a DNA polymerase III. The short RNA primer that begins the synthesis of each okazaki fragment of the lagging strands are assembled by a type of RNA polymerase known as primase. The short segments of RNA (primer) at 5′ end of the leading strand and at 5′ end of okazaki fragment are later removed and the resulting gap in the strand is filled by DNA polymerase I and then sealed by DNA ligase enzyme (Fig. 3.7).

The process of primer requirement in DNA synthesis is amazing. One of the main reason for primer requirement to initiate DNA synthesis is to avoid misincorporation of wrong nucleotides at the initial stage of synthesis. Any mistake committed due to incorporation of wrong nucleotide during DNA strand synthesis could be accurately eliminated by DNA polymerase I. Therefore fedility of DNA synthesis of strictly monitered by DNA polymerase.

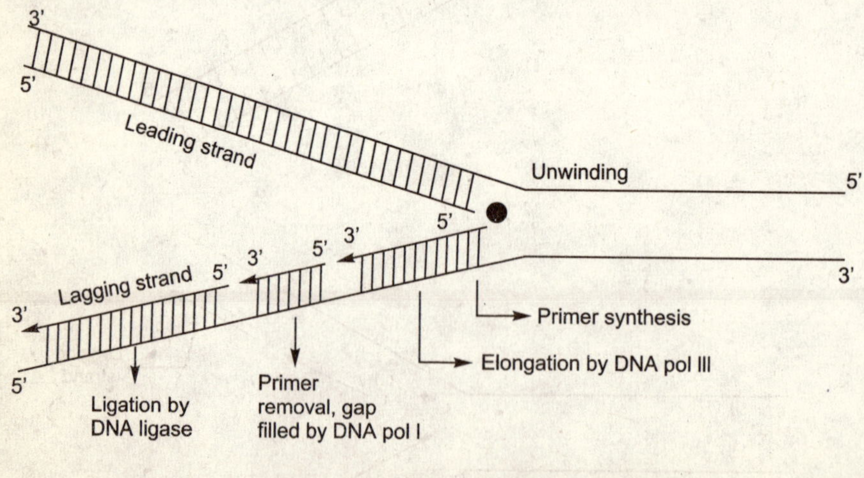

Fig. 3.7. *Synthesis of leading and lagging strand.*

3.5.1 Fedility of Replication

Fedility of DNA replication is strictly controlled by the involvement of DNA polymerase II. Survival of an organism is dependent upon the accurate replication of genome. Any mistakes committed during replication result in permanent mutation and would be detrimental for the cell. It is estimated that one out of one billion nucleotides are not complementary to the corresponding nucleotide on the template strand. Incorporation of any mismatch nucleotide will be removed immediately by DNA repair enzyme i.e., DNA polymerase. Therefore accurate proofreading system is existed during replication.

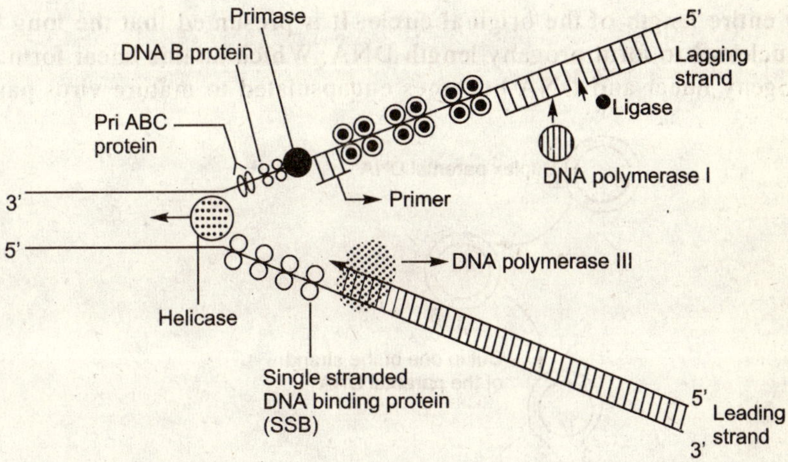

Fig. 3.8. *Involvement of proteins in DNA replication.*

3.6 TERMINATION OF REPLICATION

When replication begins at specific site (ori) on the bacterial chromosome separation of strand is followed by the entry of DNA helicase into single stranded region. The replication fork proceeds very rapidly and eventually meet in a region of the chromosome known as ter. When replication terminates there is an existence of genetically silent termination region in E. coli.

3.7 REPLICATION OF CIRCULAR DNA (*ROLLING CIRCLE MODEL*)

Replication of double-stranded circular DNA molecules can be carried out alternatively by rolling circle mechanism. Rolling circles are alternative route for replicating circular DNA. This mode represents the multiplication of many viral DNA including λ phage virus. In this mechanism DNA synthesis begins with a specific cut at origin of replication in one specific strand of the double circular DNA. The break in one of the strand releases 5′ end and it dissociate from the parental doubling DNA. As a result the free 3′ OH is exposed to DNA polymerase III function i.e., DNA pol III add deoxyribonucleotide to 3′OH region. As the replication proceeds 5′ end is rolled out gradually and is increasing in length. The active force that unwinds 5′ tail is due to the movement of the replisome helped by helicase as its key component. After the displacement of 5′ end which is soon covered by single strand DNA binding proteins. Meanwhile polymerization takes place at the 3′ end by adding new deoxyribonucleotide due to involvement of DNA polymerase. This replicating structure is called rolling model. This is accomplished by rotation of the double-helical template about its axis.

Since rolling circle mechanism involves in the replication of double stranded DNA, its displaced 5'end is soon converted into double stranded. This double stranded DNA status is contrived by the formation of okazaki fragments that extends growth from RNA primers initiated by an enzyme primase present in the primasomes. The entire polymerase III-primasome complex together called replisome continued to move and unwind 5′ tail propelled by helicase enzyme (Fig. 3.9). this elongation sometimes continue to produce long tail, sometimes the entire length of the original circle. It is presumed that the long tail are cut by specific endonucleases to form progeny length DNA. Which attains linear form. The resulting unit length progeny linear and DNA becomes encapsulated to mature virus particle.

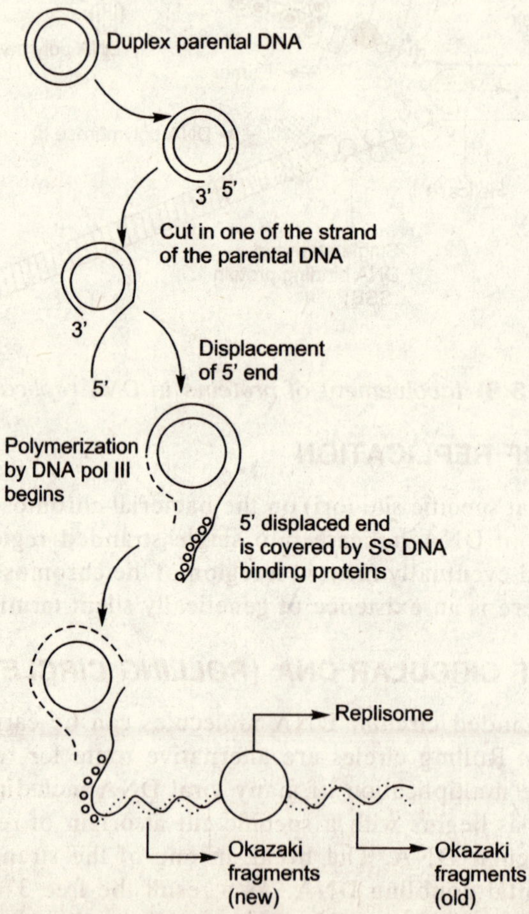

Fig. 3.9. *Rolling circle model of DNA replication.*

3.8 REPLICATION IN EUKARYOTIC CELL

Although an impressive progress has been made in the understanding of replication, but its mechanism is not yet understood completely among eukaryotic system. Initiation of replication is highly controlled and thus complicating the process. In addition, DNA is not nacked in eukaryotes but in the form of chromatin where in association of DNA with protein could be seen. The genetic material DNA is surrounded by protein called histones as a sign of better organization. Thus, higher level of organization have further complicated in understanding the process.

Replication process in eukaryotes is basically similar to prokaryotes. Replication takes place in eukaryotes due to the participation of series of proteins like unwinding proteins, SS DNA binding proteins, topoisomerase, primase, DNA polymerase and ligase. The DNA synthesis in eukaryotes follows discontinuous manner. The lagging strand in eukaryotes is much smaller than in bacteria. Their average length is about 250 nucleotide in length. There are five different DNA polymerases have been isolated from eukaryotic cell and they are designated as α, β, γ, δ and ε. Tight association between α DNA pollymerase and primase initiates synthesis of primers at 5′ region. The leading and lagging strands are assembled by polymerase δ. The role of polymjerase γ is confined to mitochondrial DNA replication. The DNA damage repair elimination of mismatch nucleotide is performed by DNA polymerase β′.

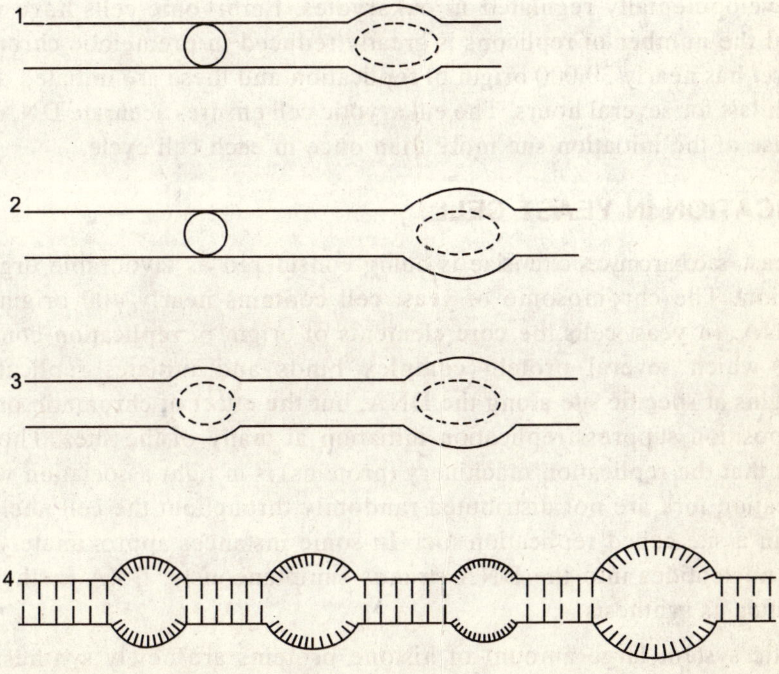

Fig. 3.10. *Multiple origin of replication in eukaryotes.*

All the eukaryotic DNA polymerase elongate DNA strand in 5′-3′ direction. These eukaryotic polymerase enzyme faithfully follows prokaryotic system by not able to initiate DNA synthesis without primer. Most of the eukaryotic DNA polymerase are able to exhibit 3′-5′ exonuclease activity to maintain highest fedility of replication.

Eukaryotic replication begins at different origins. Replication takes place simultaneously at many sites along the entire length of chromosome ensures DNA synthesis. Eukaryotic polymerases insert nucleotide growth of chain at much lower rates. The DNA is replicated in smaller units called 'replicon'. The replicons in eukaryotes are between 15 to 100 μm in length. Replication forks moves in both the direction (Fig. 3.10). DNA replication is depending upon the compact arrangement at chromatin regions. The time in which particular region of chromosome undergoes replication can be predicted. Replication is slow down at tightly compacted regions in the chromosome. Each chromosome set has 30,000 origins of replication that are activated for several hours in S phase of cell cycle.

3.9 EUKARYOTIC DNA SYNTHESIS IS BIDIRECTIONAL

Synthesis of DNA in eukaryotes is similar to prokaryotic system. Hence, prokaryotes and eukaryotes shared a common features of bidrectional synthesis of DNA. Confirmation regarding this process was clearly evidenced by autoradiography technique. The number of replicon is developmentally regulated in eukaryotes. Embryonic cells have many origin of replication and the number of replicons is greatly reduced in premeiotic chromosomes. Each chromosome set has nearly 30,000 origin of replication and these are initiated during 'S' phase duration which last for several hours. The eukaryotic cell ensures accurate DNA duplication by avoiding the use of the initiation site more than once in each cell cycle.

3.10 REPLICATION IN YEAST CELL

A common yeast saccharomyces cerevisiae is being considered as favourable organism to study DNA replication. The chromosome of yeast cell contains nearly 400 origin of replication throughout DNA. In yeast cells the core elements of origin of replication contains atleast 11 base pairs to which several protein complex binds and initiates replication. Although replication begins at specific site along the DNA, but the effect of chromatin organization and nucleosomes position suppress replication initiation at many of the sites. The Radiolabelled studies reveals that the replication machinery (proteins) is in tight association with the nuclear matrix. Replication fork are not distributed randomly throughout the cell nucleus but instead localized within a site called replication foci. In some instances approximately 40 replication incorporating nucleotides into the DNA strands simultaneously. DNA methylation progress immediately after its synthesis.

In eukaryotic system large amount of histone proteins are newly synthesized during 'S' phase of the cell cycle and are quickly associated with newly replicated DNA resulted nucleosome as a sign of highest degree of organization.

3.11 ENZYMOLOGY OF DNA REPLICATION

The process of DNA replication involves participation of several class of highly committed enzymes like nucleases, polymerases, primases and ligases. The efficacy of these committed enzyme is due to their exonuclease, endonuclease and polymerase activity.

3.11.1 DNA Polymerases

DNA polymerase enzymes are responsible for the synthesis of new strands. Investigation on these enzyme was began in 1950 by Arther Korenberg at Wasington University. However, in 1959 Koreberg discovered these polymerases that was responsible for the polymerization of deoxyribonucleoside triphosphate to construct new complementary DNA using DNA as template. Korenberg and his associates in their initial experiments were able to extract and purify enzyme from E. coli cultures. It was then named as Korenberg enzyme after its discovery. They exhibit incorporation of radiolabelled DNA precursors (nucleotides) into a DNA polymer. It was found that for initiating DNA synthesis Korenberg DNA polymerase employs four deoxyribonuleoside triphosphates such as dATP, dGTP, dTTP, dCTP and a primer along with a DNA template. In prokaryotic organisms there are three DNA polymerase activity representated as DNA pol I, II and III. In eukaryotes there are three polymerases represented as α, β and γ. The DNA polymerases of prokaryote and eukaryotes perform similar functions. DNA polymerases cannot initiate DNA synthesis. It is capable of adding nucleotides to the 3′ hydroxyl end of an existing strand. They require specifically a primer. Which provide 3′ OH terminues to the enzyme to initiate DNA strand. All DNA polymerase require template strand to copy and primer strand to initiate new DNA strand. Sometimes these enzymes are called replicases. Since these are involved in replication, others (polymerase I and II) are participate in DNA repair to replace damaged sequence.

3.11.2 DNA Polymerase I (pol I)

The DNA pol I is a single polypeptide chain with a molecular weight of 103 KD. It is associated with a single Zn atom. It is one of the most distinct polymerase present in bacterial cells. There are 300 to 400 molecules of DNA polymerase I per bacterial cell. Pol I of E. coli exhibit three important activity, i.e., polymerization, exonuclease activity (3′–5′) and exonuclease activity in 5′–3′ direction.

Although DNA pol III is the main polymerization enzyme. DNA pol I can polymerize nucleotide at the rate of 1000 molecules per minute. E. coli can initiate replication process in the absence of pol I. This suggested pol I enzyme do not effect DNA synthesis and rather it can concentrate on synthesizing short stretch of DNA fragment. It is even participate in repair synthesis. The polymerase activity of this enzyme is in 5′-3′ direction. DNA-pol I exhibit 3′-5′ exonuclease activity as it catalyses the breakdown of one of the strand of DNA into it nucleotide in 3′–5′ direction. One of the most outstanding contribution of pol I in E. coli cells is that it can perform proof reading and remove mismatched nucleotide i.e., addition of any wrong nucleotide will be removed immediately. Such an activity play a significant role in maintaining accuracy of DNA synthesis. Removal of nucleotide can be done by pol I even in 5′–3′ direction.

Ultraviolet can damage DNA to the greater extent depending upon the wave length of the UV exposure and sensitivity of the cells. Thimine dimmer occur when cell exposed to ultraviolet Presnece of dimer can jeopardize the fate of DNA especially during its replication. Hence, polymerase I take up special task of removal of thiamine dimer (T = T) from the DNA strand, this suggest that this enzyme involved in repair synthesis. During DNA synthesis small-fragments known as okazaki fragments are synthesized, to each of which attach a small RNA primer. The action of primase leaves a short stretch, if RNA at 5′ end of each okazaki fragments. These are removed in the later stage by 5′–3′ exonuclease activity of pol I at 5′ ends of each okazaki fragments. The gap is filled immediately. Thus, replacement of DNA is accomplished by the same enzyme. DNA polymerase can be cleaved into regions. It comprises of larger fragments and smaller fragments. Enzyme trypsin can split this into different length of the fragments. The larger cleavage fragments are called Klenow fragments (68 KD). The larger fragments exhibit 3′–5′ exonuclease activity besides its polymerization strength. While smaller fragments can show 5′–3′ exonuclease activity. Synthesis as well as degradation properties of DNA pol I was well characterized in E. coli. In addition these activity (polymersiation, 3′–5′ exonuclease) can be found in different domains of the single polypeptide.

3.11.3 DNA Polymerase II (pol II)

There are about 40 molecules per cell. It consist of single polypeptide chain with molecular weight of 90,000. The presence of pol II is marked by the presence of highly concentration of DNA pol I. It shows 5′–3′ exonuclease activity. The polymerase potential is less than that of pol I and it can incorporate only 40-45 nucleotide per minute. This enzyme do not show 5′–3′ exonuclease activity. Unlike in DNA pol I pol II 5′–3′ exonuclease suggest that it can perform repair synthesis of DNA damaged by external agent like UV. In the absence of pol I, this enzyme can shoulder the responsibility of elongating okazaki fragments. Therefore, pol II can be an suitable alternative candidate for pol II.

3.11.4 DNA Polymerase III (pol III)

DNA polymerase III is called as replicase. It is a large enzyme complex of 130 KD α subunit. The presence of this subunit is responsible for DNA synthesis activity. The proofreading and 3′–5′ exonuclease job is performed by another subunit ′ε coded by a gene DNA Q. This enzyme exhibit 5′–3′ polymerisation activity and 5′–3′ exonuclease activity and proof reading activity. They also require primer to initiate DNA synthesis. It exhibit greater degree of activity and has as definite edge over pol I and II. It is basically a dimer of two polypeptide chain with a molecular weight of 1,40,000 and 40,000. Polymerisation potential is very high in 5′–3′ direction. It can add 1,50,000 nucleotide per minute during polymerization process. In addition pol III also shows 3′–5′ exonuclease activity. One of the most promising functions of pol III as its dual nature of both 3′–5′ and 5′–3′-exonuclease activity.

3.12 CYCLING MOVEMENT OF DNA POL III

The model on movement of DNA pol III have shown in several experimental evidence in which the polymerase bind to the DNA by the β sliding clamp (Fig. 3.11)

As it moves along the template strand it can synthesis complementary strand. After the completion of okazaki fragment synthesis. The enzyme dissociate from β clamp and acquire a cyclic move to a new β clamp on a upstream RNA primer. Meanwhile the original β clamp is left behind on the completed okazaki fragments.

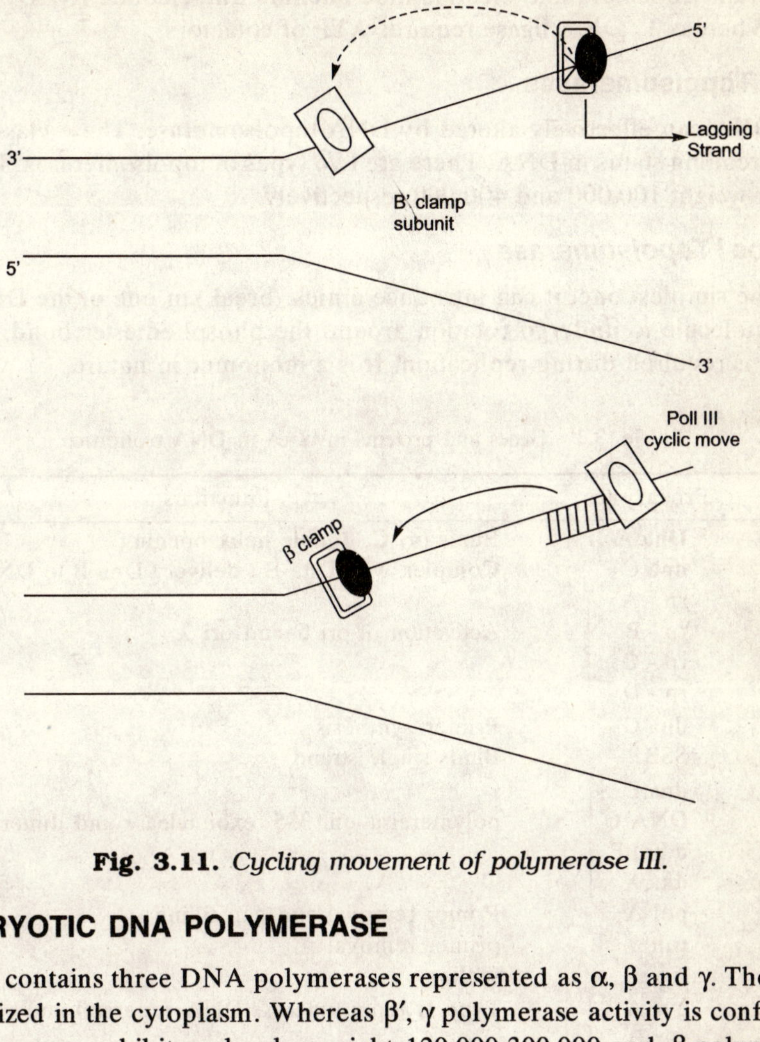

Fig. 3.11. *Cycling movement of polymerase III.*

3.13 EUKARYOTIC DNA POLYMERASE

Eukaryotic cell contains three DNA polymerases represented as α, β and γ. The α-polymerase activity is localized in the cytoplasm. Whereas β′, γ polymerase activity is confined within the nucleus a polymerase exhibit molecular weight 120,000-300,000 and β-polymerase measure about 40,000 similarly γ-polymerase shows 250,000. Mammalian cell organelle like mitochondria also shows DNA polymerase activity. It's a and γ polymerase have a molecular weight of over 100,000 while β polymerase can reach 30,000. In general, all the basic properties of DNA polymerases are similar to prokaryotic DNA polymerase.

3.13.1 Primases

Primases are RNA polymerases which involve in the synthesis of a short stretch of primer on the template DNA to facilitate DNA synthesis.

3.13.2 DNA Ligase

DNA ligases are otherwise known as "Molecular Switcher". It can seal single strand DNA nick in DNA. It catalyses the formation of phosphodiester bond between 3'OH and 5'PO_4 group. DNA ligases is of two types one is E. coli based DNA ligases and other one ; viral based T_4 DNA ligase certain co-factors like nicotinamide adenine dinucleotide (NAD) are required by E. coli ligase. Whereas T_4 DNA ligase requires ATP of cofactors.

3.13.3 DNA Topoisomerases

Shape of the DNA can effectively altered by DNA topoisomerase. These class of enzyme can introduce supercoiling status in DNA. There are two types of topoisomerases. Type I and Type II of molecular weight 100,000 and 400,000 respectively.

3.13.3.1 *Type I Topoisomerase*

The type I is the simplest one, it can introduce a nick (break) in one of the DNA strand. This results in the molecule to undergo rotation around the phosphodiester bond on the opposite strand, since it is required during replication. It is a monomer in nature.

Table 3.1 Genes and proteins involved in DNA replication

Genes	Prokaryotes	Functions
dna A	Dna A	Binds ori C, double helix opening
dna C	dna C	Complex with Dna B - delivery Dna B to DNA
RNA polymerase	rp - A rp - B rp - C rp - D	Activation of ori C and ori λ
Primase	dna G	Primer synthesis
SSB	SSB	Binds single strand
DNA poly merase III	dna E DNA Q d hol E dna X	polymeraisation, 3'-5' exonuclease and dimerisation of core
pol I	pol A	Primer removal and gap filling
RNase H	rnh	primer removal
Ligase	lig	sealing
Gyrase	gyr A gyr B	Type II topoisomerase-DNA supercoil and replication swivel
TBP	tus	Ter-binding
Helicase II		Helicase

3.13.3.2 *Type II Topoisomerase*

Type II DNA topoisomerase do not require ATP for their activity. They bind covalently to the cut strand and provide sufficient energy to the phosphodiester bridge. The acquired energy is in turn utilized during sealing of the nicked strand after the tension is released from the DNA. The gene for type II is identified as gyr A (Table 3.1)

Type II topoisomerase can invade and cut both the strand and have capacity to seal the strands. They can relieve tension in the DNA (Fig. 3.12). This require ATP energy to perform the functions of negative supercoiling. Where it is ATP independent for DNA relaxation. Eukaryotic topoisomerase have also contains type I and type topoisomerase of the molecular weight 100,000 and 309,000. Type II is dimer and ATP independent of DNA relaxation.

3.14 REGULATION OF DNA REPLICATION

Regulation of DNA synthesis in various cell types is an enigma. Several cell types are well equipped with regulatory mechanism of DNA synthesis at various level. One of the simple method is to restrict DNA replication to limited period within the cell cycle and avoid DNA replication to limited periods within the cell cycle and avoid the use of given origin for replication only once in a given cycle of chromosome replication. The restricted use of origin of replication site is successfully achieved by the involvement of the origin site cannot be used again as a starting site for DNA synthesis. Although it is believed that methylation, demethylation take place at the modified site of DNA by dna A protein but there is no evidence on the exact chemical modification at replication site.

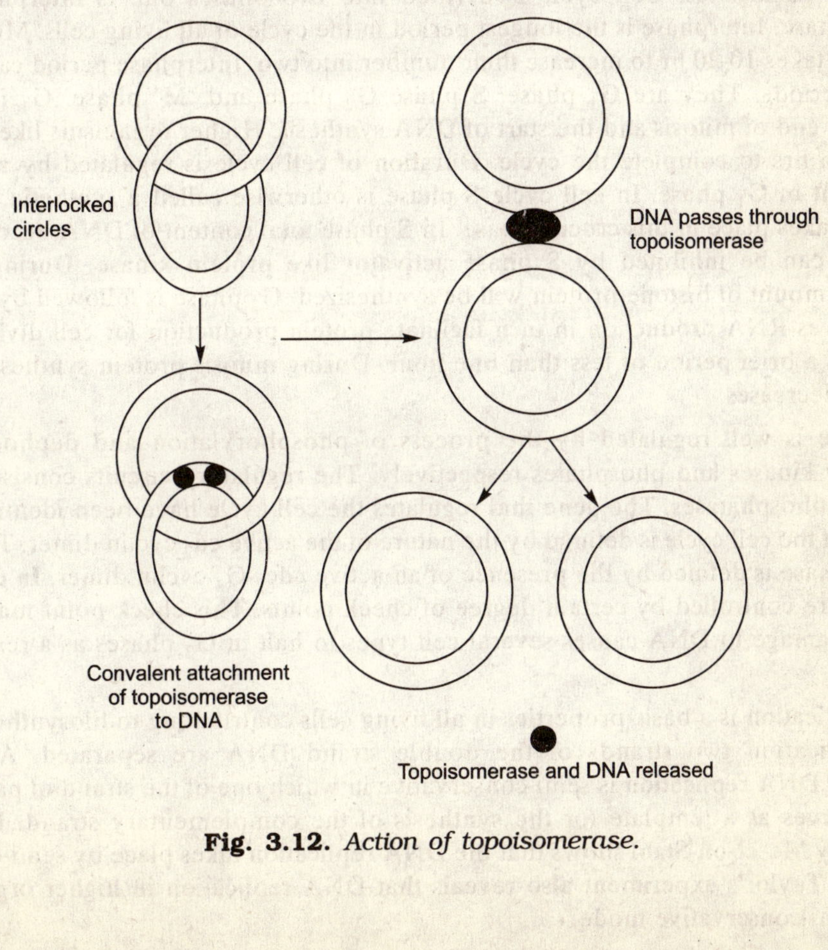

Fig. 3.12. *Action of topoisomerase.*

Table 3.2 Comparison of Prokaryotic and Eukaryotic DNA Replication

	Prokaryotic	Eukaryotic
1. Number of origin of replication	01	Thousands in number in the chromosome
2. Okazaki fragments and lagging strands	1000-2000 nucleotides	less than 200 nucleotide
3. RNA primer length	50 nucleotide	9 nucleotide
4. DNA polymerisation rate (fork movement rate)	500 nucleotide per second	50 nucleotide per second
5. Cell cycle 'S' phase	Not differentiated	Cell is involved replication takes place in

SUMMARY

Cell cycles proceeds in all eukaryotic organism. Cell cycle is the period between two mitotic divisions in eukaryotes. Cell cycle is divided into two phases one is interphase other is divisional phase. Interphase is the longest period in life cycle of all living cells. Most plant and animal cells takes 10-20 hr to increase their number into two. Interphase period can be divided into four periods. They are G_1 phase, S phase G_2 phase and 'M' phase. G_1 is the period between the end of mitosis and the start of DNA synthesis. Higher organisms like mammalian cell takes 16 hrs to complete the cycle. Duration of cell cycle is regulated by arresting at a specific point of G_1 phase. In cell cycle S phase is otherwise called a synthetic phase. DNA replication takes place in this crucial phase. In S phase total content of DNA becomes double. This phase can be inhibited by S phase activator like protein kinase. During this phase substantial amount of histone protein will be synthesized. G_2 phase is followed by S phase. G_2 phase involves RNA production in turn facilitate protein production for cell division. Finally 'M' phase is a brief period of less than one hour. During mitosis protein synthesis and cyclic Amp level decreases.

Cell cycle is well regulated by the process of phosphorylation and dephosphorylation catalysed by kinases and phosphates respectively. The regulatory circuits consist of series of kinases and phosphatases. The gene that regulates the cell cycle have been identified as cdc_2. The phase of the cell cycle is defined by the nature of the active cd_2 cyclin dimer. The period of G_1 and S_1 phase is defined by the presence of an active cdc_2-G_4 cyclin dimer. In cell cycle, all the phases are controlled by certain degree of check points. This check point make a control loop. Any damage to DNA causes several cell types to halt in G_2 phases as a result of check point.

DNA replication is a basic properties in all living cells contributing to biosynthesis of DNA. During replication two strands of the double strand DNA are separated. According to investigators DNA replication is semi-conservative in which one of the strand of parental DNA molecule serves as a template for the synthesis of the complementary strand. Experiments conducted by Meselson-Stahl shows that the DNA replication takes place by semi-conservative mechanism. Taylor's experiment also reveals that DNA replication in higher organism takes place by semi-conservative mode.

DNA replication takes place due to the involvement of committed enzymes and proteins. Replication begins with recognition of origin of replication. Which is followed by strand separation and attachment of RNA primer.

The first step in the initiation of replication is the specific binding of dna A protein to oric. This will open the double helix. Strands separation is made by dna B gene products. Meanwhile topoisomerase released highly tensioned supercoiled DNA into normal DNA. Unwinding of DNA is done by DNA helicases. Once replication starts, replication proceeds bidirectionally along the entire length of DNA. Replication fork can be seen at the site where the parental strand is undergoing strand separation and nucleotides are assembled.

The first step in the replication of DNA is the synthesis of RNA primer. These are 50-100 nucleotide long synthesized by primase. DNA polymerase cannot initiate synthesis of DNA on its own capacity. However, it can initiate synthesis on the primers when they are available. DNA replication will be carried out smoothly by the participation of three DNA polymerases. The DNA pol III is responsible for the synthesis of two DNA strands at the replication for the synthesis DNA in 3′–5′ direction. DNA pol I facilitates DNA repair during the synthesis of new strand. The DNA pol II is probably synthesise DNA in 3′–5′ direction.

It has been shown that replication is continuous on one strand and discontinuous on the other. DNA strand cannot be synthesized in 3′–5′ direction rather both the newly synthesized DNA strands proceeds 5′-3′ direction. One of the DNA strand can be synthesized by DNA pol III in 5′–3′ direction using 5′–3′ DNA as template. However DNA pol III cannot synthesise DNA is antiparallel 5′–3′ strand. The strand that synthesizes away from the replication fork is synthesized discontinuously like fragments. Hence the strands synthesise continuously is called the leading strand and it is synthesised discontinuously in opposite strand is called lagging strand. The small fragment of DNA is known as okazaki fragments discovered by okazaki of Nagoya University Japan. The leading strand of DNA is initiated by RNA polymerase. Once this primer is synthesised, which is then extended by DNA polymerase III. Primers are later removed by DNA pol I and then sealed by ligase.

Fedility of DNA replication is strictly controlled by the involvement of DNA polymerases II. Incorporation of wrong nucleotide would be detrimental to the cells but those wrong nucleotides are eliminated immediately by polymerase enzyme.

In bacteria replication of double stranded circular DNA molecules can be carried out alternatively by rolling circle mechanism. This mode represents even the multiplication of λ phage virus.

In eukaryotic DNA replication one can see several hundreds of origin of replication throughout the chromosome. There are five different DNA polymerases have been identified. They are α, β, γ, δ and ε, involved in replication. All the eukaryotic DNA polymerase elongate DNA strand in 3′–5′ direction. Eukaryotic replication begins at different origins simultaneously. Thus, several replication forks can be seen during DNA replication. Eukaryotic DNA synthesis in bi-directional like prokaryotes. In eukaryotes once the DNA synthesis is completed it is immediately associated by histone proteins and organized into nucleosomes.

Several cell types are well equipped with regulatory mechanism of DNA synthesis at various levels. One of the important method is to restrict DNA replication to limited period within the cell cycle and avoid use of replication only once in a given cycle.

Both prokaryotic and eukaryotic DNA replication is proceeded by the active participation several committed proteins and enzymes. These are topoisomerases, helicases, three types of DNA polymerases in prokaryotes, three polymerases in eukaryotes. The enzyme gyrases or topoisomerases are of two types can interconvert super coiled into relaxed DNA and vice versa. As the helicases enzymes unwinds DNA helix, single stranded (SS) DNA binding proteins binds to the DNA to prevent reassociation of helix. DNA polymerases can perform exonuclease activity in both the direction, one of the most important enzyme during replication is DNA polymerase III called replicase, exhibit 5′-3′ polymerisation activity and also proof reading activity. However, they require primer to synthesise DNA. DNA pol I can further complete the synthetic process and DNA ligase finally seals the pieces DNA (okazaki fragments).

QUESTIONS

1. Enumerate the differences that distinguish eukaryotic from prokaryotic replication. (BU May, 1996)
2. List out the different enzymes involved in DNA replication. Add a note on their function. (BU Oct/Nov., 1998)
3. Describe the process of replication in prokaryotes. (BU Nov., 1998)
3. Describe the process of replication in prokaryotes. (BU Nov., 1999)
4. Explain the DNA replication in Eukarotes. (BU Nov., 1999)
5. Describe the process of Prokaryotic and Eukaryotic replication. (BU April 2000)
6. Enzymes play an important role in DNA replication justify. (BU April 2001)

SHORT QUESTIONS

1. Okazaki fragments (BU April, 1998) (BU May, 2002)
2. DNA polymerase (BU April, 1998) (BU April, 2001)
3. DNA ligase (BU Oct., 1998)
4. Eukaryotic DNA polymerases (BU Nov., 1999) (BU April 2000)
5. Bacterial RNA polymerase (BU Nov., 1999)
6. DNA binding proteins (BU April, 2000)

4
CHAPTER

The DNA Repair Mechanism

Although DNA molecule inside the cell is completely safe, it is however subjected to damage. Factors like ionizing radiation and ultraviolet are some of the biggest threat to the master molecule DNA. Experiments with bacteria reveal that the cell has a remarkable ability to repair such damage. This giant molecule is replicated with extraordinary fedility in a cell and placed DNA in the safest position. The cellular repair machinery is committed to safe guard this hereditary molecule. The DNA repair enzymes is swung into action if they see even a minor damage to this molecule. The DNA is one of the molecule in a cell that is most susceptible to environmental damage. The backbone of the DNA molecule is broken when it is subjected for ionizing radiation. Some of the cell's own metabolic product can also damage DNA, for example, free radicals can severely damage DNA. By damaging DNA these free radicals can cause harm to the greatest extent. Until recently it had been thought that if the DNA in a cell were damaged or altered by ionizing radiation, the cell might have given rise to mutant daughter cell or no daughter cell at all. Now it is an established fact that the cells are equipped with potential repair mechanism. Ability of cells to repair damaged DNA may have been a significant factor in the biological evolution. In addition, the repair would be advantageous in species to maintain high stability in the environment.

Detection of errors and immediate correction mechanism is exemplary. If a portion of one strand of the DNA helix were damaged information in that portion could be obtained from the complementary strand, that is, the cell could use the undamaged strand of DNA as a template for the reconstruction of the damaged segment in the complementary strand. In fact, biologists have realized that cells have a variety of repair system that are able to correct any damage to which they might be susceptible. It is estimated that only one base out of thousands can escape cell repair system. But prokaryotic and eukaryotic cells exhibit a variety of enzymes that patrol DNA molecule. In some cases, damage can be repaired directly. One of the classic example is a type of enzyme that utilizes the energy of sunlight to break the bond holding two pyrimidine together in a pyrimidine dimer.

4.1 PROOFREADING IN DNA REPAIR

Maintenance of correct base sequence during replication has been completely understood in E. coli as a model system. All living cells enjoy proofreading system using DNA polymerases.

The error frequency in E. coli DNA replication is incorrect base in 10^4 nucleotides. Presence of 10 wrong nucleotides in 10^4 base pairs could be still potentialy harmful to the cell when it undergoes next replication cycle. But the amount of mutation recovery may be negligible.

The enigma of replication accuracy was cleared when proofreading system was discovered. When 10 wrong bases were given to DNA polymerase I and III, the incorporated base will be rectified immediately by the removal of wrong base by 3′-5′ exonuclease activity of DNA pol I. Therefore, it was concluded that if any wrong base is accidentally incorporated, it is eliminated immediately by proofreading mechanism. In eukaryotes polymerase δ shows 5′-3′-exonuclease activity to remove wrong bases.

4.2 TYPES OF DNA DAMAGE

As it was mentioned earlier ultraviolet radiation from sunlight and X-ray irradiation are the two main culprits that can cause damage to DNA. The ultraviolet can cause dimer formation between two adjacent thiamine residues in the DNA molecule. These dimerisation produces thiamine dimer. They could potentially damage DNA unless they are removed from the DNA molecule. Certain reactive chemicals such as alkylating agents like nitrosogunaidine and nitrosoamine inflict damage by modifying base pair and allow wrong base pairing during DNA synthesis.

In addition, several mechanical forces generated within a chromosome like bending or shear force result in DNA strand break. Sometimes, the change in local pH may result in purine loss and contribute to DNA damage.

4.2.1. UV Radiation and DNA

DNA strongly absorbs ultraviolet primarily by the bases. After studies on the effect of UV on the isolated base, it was noted that thymine and cytosine belonging to the group pyrimidines are far more sensitive to ultraviolet than purines. In 1960, Adolf Walker at the University of Frankfurt found thiamine dimer in DNA extracted from irradiated bacteria. Thus, thiamine dimer is a formation of additional bond between two adjacent thiamine bases on DNA molecule. Julius and Lawrence of Brande's University have shown that when a double strand DNA is exposed to ultra violet the two strands become more strongly linked by chemical bond rather than original weak hydrogen bond. It was evidenced that strong link between two strands is the result of formation of dimers between thiamine units on opposite strands of the double helix. Presence of dimer in DNA molecule would interrupt the separation and may disrupt replication process. Dimer between adjacent thymines on the same strand would interfere with proper pairing of the bases. The joining of two adjacent thymines would alter incorporation of adenine. Therefore replication might stop at that point or proceed incorrectly with altered base sequence on the newly formed DNA chain (Fig. 4.1).

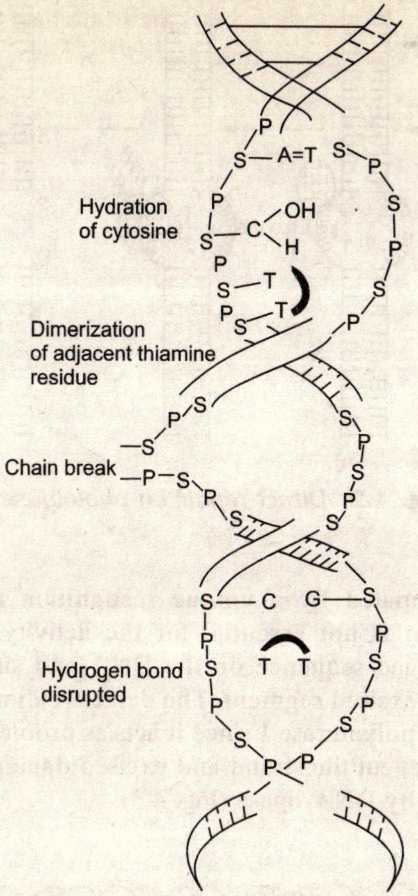

Fig. 4.1. *Alteration of DNA by UV irradiation.*

4.3 TYPES OF DNA REPAIR

Almost all the living cells contain certain degree of enzymes involved in repairing of damaged DNA. There are four important type of DNA damage repair system exist as follows.

4.3.1. Photoreactivation or Dimer Repair

This type of repair involves reversal or simple removal of damage. The DNA repair is done by light dependent enzyme known as photolyase. It can reverses or remove dimer caused by ultraviolet. This enzyme is wide spread in nature especially in plant system. In E. coli single gene product (phr) codes for an enzyme photolyase. It simply binds on the defective dimer site when light falls on it and the enzyme absorbs light energy. When enzyme molecule get energized. It would promote cleavage the bond at dimer site in one of the strand (Fig. 4.2).

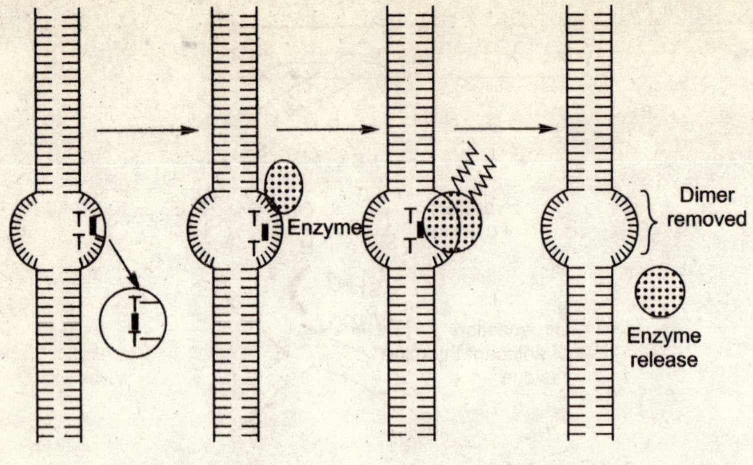

Fig. 4.2. *Direct repair by photolyase.*

4.3.2. Excision Repair

This repair mechanism is initiated by a unique recognition of enzyme. This process is independent of light i.e., light is not essential for the activity of enzyme. Recognition is followed by removal of damaged sequence on the DNA and simultaneously new stretch of DNA is synthesized to replace excised segment. The defective dimer region is detected by 5′-3′ endonuclease activity of DNA polymerase I since it acts as proofreading or editing function. It can recognize the defective site, cut the strand and excised damage dimer site. After synthesis of new strand it is then sealed by DNA ligase (Fig. 4.3).

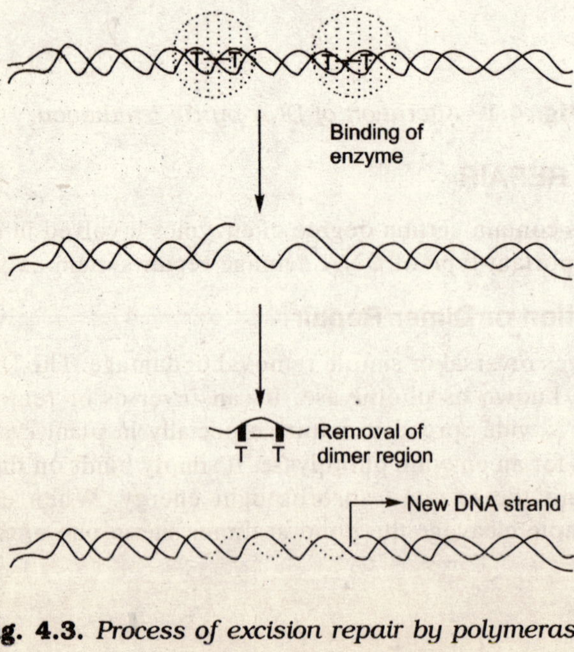

Fig. 4.3. *Process of excision repair by polymerases.*

4.3.3 Excision Repair by Rescue Enzyme

Some of the mutation is caused by alkylation of O^6 of guanine. Once DNA is methylated certain enzymes like methyl guanine methyl transferase ready to strike methylated part and comes for repair. Other enzymes like DNA glycosylase, which breaks the glycosidic bond between the damaged base (caused by mutation) and its sugar. This would create apyrimidine site popularly known as AP site. This site is recognized by 5′ AP endonuclease that break or nick the DNA strand. In addition 5′-3′-exonuclease also start at the nick and move left to right. This DNA glycosylase enzyme can also remove pyrimidine dimer caused by UV light.

4.3.4 Recombination Repair

If a DNA template for repair synthesis is not available. An alternative recombination method is suitable to overcome DNA damage. In this process large segment of damaged DNA is completely replaced by corresponding segment of its partner DNA i.e., homology duplex of DNA. Studies of E. coli shows that the enzyme ReCA protein can favour DNA strand exchange and promote recombinational repair. It can even mediate the induction of 15 DNA repair genes called SOS genes.

4.3.5 Retrieval Repair System

When damaged DNA sequence remaining in a daughter molecule. The replication of DNA can be forced to bypass at damaged site and retrieval system uses recombination method to obtain another copy of the sequence from undamaged segment of DNA. This has been well characterized in several bacteria.

Several people are sensitive to ultraviolet component of sunlight may result in skin reaction, some time even leads to skin cancer (Table 4.1).

Table 4.1 Effect of UV on human being

Disease	Sensitivity	Symptoms
1. Xerodema pigmentosa	UV irradiation alkylation	photosensitivity
2. Bloom's syndrome	UV irradiation	Photosensitivity facial changes
3. Fancon's anaemia	cross linking agent	General decrease in number of blood cells may leads to leukamia
4. Ataxia felangiectasia	γ radiation	Dilation of blood vessels in skin and eye, chromosomal abberation

4.3.6 SOS Repair

When DNA undergoes severe damage, in respone to it an SOS repair system is induced. The SOS repair system is highly complex system involving 20 genes. Two of these regulatory genes involved in SOS repairs system are LEXA and RECA.

Several DNA damaging process can activate RECA during SOS response. The inducing signal could consist of a small molecule released from DNA or it might be relevant structure formed in the DNA itself. RECA activation is aided by presence of single-stranded DNA and ATP. Thus the activation signal may be the presence of a single stranded region at the site of damage. There have been several reports confirms that SOS-response swung into action within few minutes of the damage. LeXA gene incodes a small 22 KD protein which undergoes cleavage the activation of ReCA.

4.3.6.1 *Repair Mechanism :*

LEXA is a dimer its monomer contains DNA binding domain. LEXA binds to its operators to block expression or the SOS repair genes. The ReCA protein has single stranded DNA binding activity. It also exhibit multifunctional characters. The ReCA co-protest (after bound by SS DNA it is co-protein) activity stimulates a protein activity to LEXA protein when inducing signal is removed. The ReCA protein loses the ability to destabilize LeXA. At this junctures, the LeXA gene is being expressed at a high level in the absence of activated ReCA. The LeXA protein accumulate swiftly in uncleaved status and switch off the SOS genes. The auto cleavage of LeXA separates the DNA binding domain from K domain. Consequently LeXA is not in a position to repress transcription and finally SOS genes are expressed (Fig. 4.4)

When SOS response takes place at DNA damaged site, many proteins responsible for repair recombination of replications are synthesized. Two of that proteins are UMUC and UMUD genes facilitates DNA replication to proceed across damaged site of the template strand. This repair system eliminates gaps in the newly synthesized strands opposite to damaged nucleotides in the template strands.

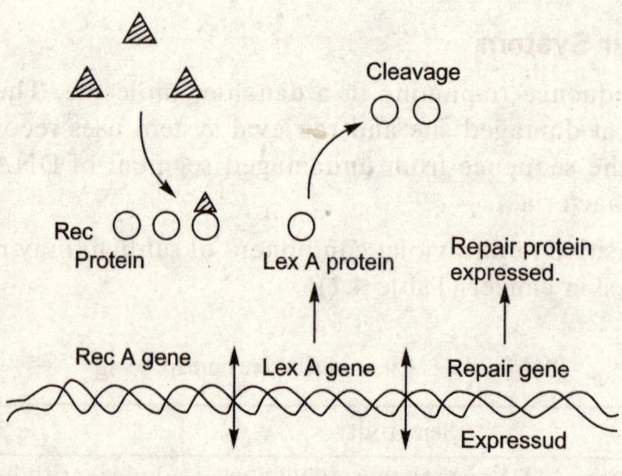

Fig. 4.4. *Gene expression during SOS repair.*

After the expression of SOS response genes. Some of their products assemble at the damaged site to form a specialized replication system depends on DNA polymerase II for replicating past DNA lession (damaged region).

SUMMARY

The genetic material DNA can be damaged by variety of factors including UV light, ionizing radiation like X ray and γ ray etc. All living cells including both prokaryotes to eukaryotes consist of potential repair mechanism. Which involves participation of several group of enzymes. Free radicals can even damage organelle DNA may leads to ageing of a cell. Several mutants can even be detrimental to the cells. Repair system can detect errors and immediately involved in the elimination of damaged segment of DNA.

Generally damage is caused by the presence of dimer. It is an additional bound formed between two thiamine bases, this dimer is removed in the cell by repair mechanism. Several repair mechanisms existed in the cells have been clearly established. Most prominents are the involvement enzymes like photolyase. DNA polymerase I and III. Apart from this DNA damage is also caused by alkylation of bases this is removed in the cells by enzyme methyl guanine. Methyl transferase can strike alkylated region and rescue DNA.

QUESTIONS

1. Explain one method of DNA repair mechanism. (BU May 1990)
2. What are the different repair mechanisms for DNA ? Explain any two in detail. (BU Nov. 1999)

SHORT QUESTIONS

1. Mismatch repair of DNA (Nov. 1999)

5 CHAPTER

DNA Recombination in Prokaryotes (Molecular Mechanism)

Recombination of genetic material at molecular level has been characterized well in prokaryotes as well as in eukaryotes. The DNA in all living cells are in the status of recombinant ones. The combining and modifying of DNA have taken place throughout the course of evolution in order to construct complex chromosomes. Recombination process helps in speciation and contribute significantly in evolution. Recombination between homologous chromosomes takes place not only in plants and animals but also in prokaryotes, viruses and plasmids. In addition this process could be seen even in the DNA of cell organelles like mitochondria. In eukaryotes, recombination takes place more precisely during meiosis. In all the cells recombination takes places so accurately in such a way that not even a single base pair is added to or lost from the recombinant chromosomes. Genetic recombination is not accidental but is an essential process in the cell catalysed by certain key enzyme that cell encode for this protein to regulate the process. These enzyme contribute significantly in accelerating genetic variation may lead to speciation in nature. One of the most important discovery made in molecular genetics is that the movable DNA segment called jumping genes or transposons, which are occasionally jump within or to another chromosomes. Thus fundamentally alter the structure of chromosomes at the site of insertion. Recombination process can even allow replacement of damaged segment of DNA during DNA repair mechanism. Generally three types of recombination occur involving physical exchange of material between duplex DNA. These process are categorized as site specific recombination, homologous recombination are non-homologous recombination.

The event of genetic recombination is possible when genetic material move from one bacterium to another. There are three different processes are involved in genetic transfer in bacteria. They are (1) Transformation (2) Transduction and (3) Conjugation. These are described in detail as follows.

5.1 TRANSFORMATION

The overall process of transformation has been well characterized due to the contribution of Griffith, Avery and McCarty in 1928 and 1944 respectively. Transformation is a natural (or induced) process in which bacteria uptake nacked DNA in to its cell. DNA uptake is energy required process. It does not involve passive entry of DNA molecule through cell the wall and membrane except under in vitro experiments. Thus, transformation cannot be seen in all species of bacteria. However, only cells exhibit window of competence can uptake of DNA. Hence, only a competent cells which possess competence factors like (probably) proteins or enzymes involved in binding or in uptake of DNA. The model on DNA uptake was proposed by Lacks in 1962. The process of transformation begins with reversible binding of double stranded DNA molecule to receptor site on the bacterial cell surface. It is followed by irreversible uptake of the donor DNA. At this stage DNA is characterized by acquiring resistance to DNase. Then double strand donor DNA is gradually converted into single stranded molecule. The DNA strand conversion is aided by enzyme. A membrane bound exonuclease or DNA translocase pulls one strand of donor DNA into the cell. The energy for pulling the DNA strand is obtained from degradation of the complementary strand of DNA. Once the donor DNA enters inside bacterial cell integration (Insertion) of all or part of the single strand DNA into the chromosome of the recipient. In the final stage segregation and expression of the integrated gene in the recombinant cell. The integration or DNA recombination is specific for homologous DNA. Thus, only the DNA of closely related species lead to recombination. During integration, a single strand of donor DNA is physically inserted into the recipient chromosome. Integration result in the replacement of one of the strand of recipient. Invitro experiments have shown the uptake of 20,000 nucleotide pairs (Fig. 5.1).

In the invitro experiments it is possible to make non-competent cells to become competent to uptake DNA. In the example E. coli cells do not possess ability of taking up functionally active DNA. It can be modified by cell surface to render more competent. The main strategy is treating the cells with $CaCl_2$. It has been shown that DNA added in presence of high level of Ca^{++} can able to transform efficiently. This method is widely employed in chemical gene transfer method.

5.2 TRANSDUCTION

Another event of genetic recombination takes place when a bacteriophage particle carries a segment of the chromosome from one bactereium to another bacterium. In this case only small segment of DNA is transferred. There are two types of transductions (a) generalized transduction and (b) specialized transduction.

5.2.1. Generalised Transduction

Based on the type of interaction between bacteria and bacteriophage. These have been classified into two types one of which is virulent phages always multiply and lyse to host cell after infecting another is temperate phages which can undergo two life cycles depending on its choice between the two. They can enter lytic cycle, reproduce and lyse the host bacterial cell immediately. It proceeds in the same pathway as virulent phages does in their life cycle.

Alternatively they can enter a lysogenic pathway, during which viral chromosomal DNA are integrated (inserted) into the chromosomes of the host. This alien guest DNA replicate like any other segments of the host chromosomes.

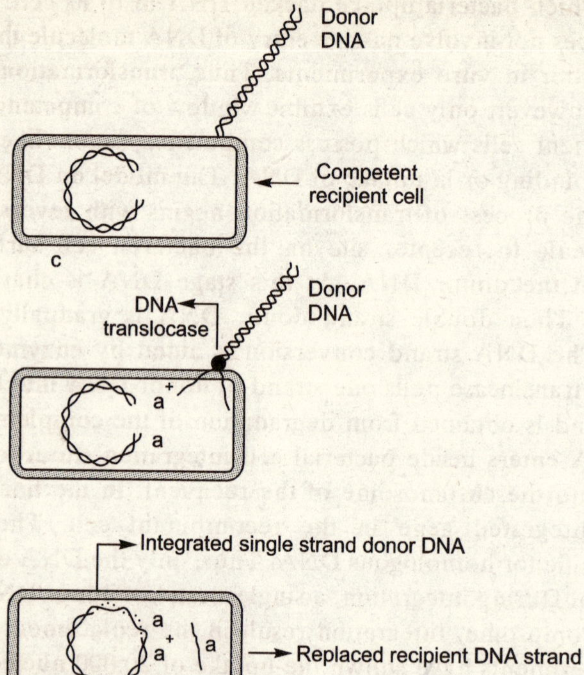

Fig. 5.1. *Transformation in bacteria.*

In generalised transduction a segment of bacterial DNA (host DNA) is accidentally wrapped up during phage maturation. Thus, certain viruses can package fragments of host DNA. Only few numbers of virulent phages known to mediate transduction.

5.2.2 Specialised Transduction (Molecular Mechanism of Site Specific Recombination)

Specialised transduction is mediated by temperate bacterium phage, whose DNA are able to integrate at one or two specific site.

The site specific recombination have been well characterized in the integration of λ DNA into the chromosomal DNA of bacteria. This recombination involves breakage and reunion at specific sites. The life cycle of λ phage virus comprises of two events. One is lytic and another is lysogenic as mentioned earlier. In the lytic cycle, lamba DNA when enters the bacterial cell exist as a independent double-stranded circular molecule in the bacterium (Fig. 5.2). In lysogenic state the phage of DNA integrate (enters) into bacterial chromosome and presence as a guest (called prophage). The prophage λ DNA can integrate and exit from bacterial chromosomal DNA at specific loci, known as attachment site (att^{λ}). When att^{λ} is removed from the bacterial chromosome λ phage integrate else where less efficiently. More precisely the bacterial attachment site (att^{λ}) is called att B. Similarly the attachment site on the λ phage is known as att^{δ} site. The sequence is common to att^{ρ} and att^{B} called as core sequence and the recombination takes place at this region. Experiments have shown that during integration of λ DNA into chromosomal DNA as prophage. It becomes inserted at a specific site on the host chromosome between gal and bio operon. Recombination takes place by phage coded gene product, its integrase and a bacterial protein called integration host factor (IHF). The insertion of λ DNA is initiated by the strands which are in close contact. The integrase enzyme catalyse break at att B site in λ DNA and another break at att λ of the host chromosome and subsequent crossed reunion of double strand of the phage and host genome.

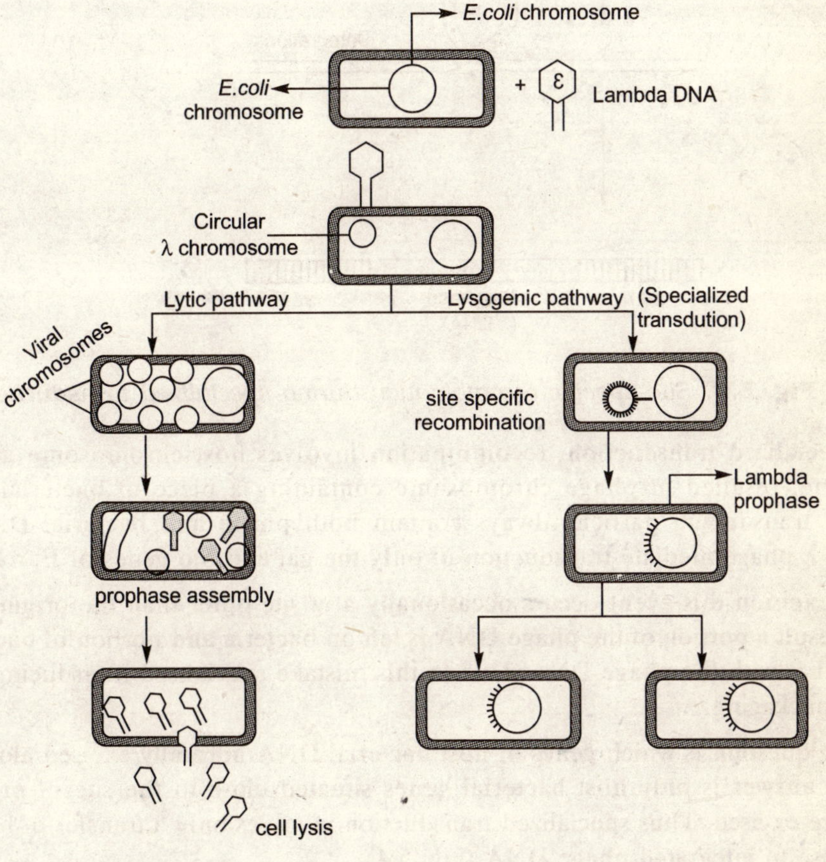

Fig. 5.2. *Lytic and Lysogenic pathway in E. coli λ phage.*

After integration into bacterial DNA. The prophage (λ DNA) is bounded by two new att sites. The products of the recombination called attL and attR. The excision requires recognition between attL and attR in addition to integrabe and integration host factor (IHF) (Fig. 5.3).

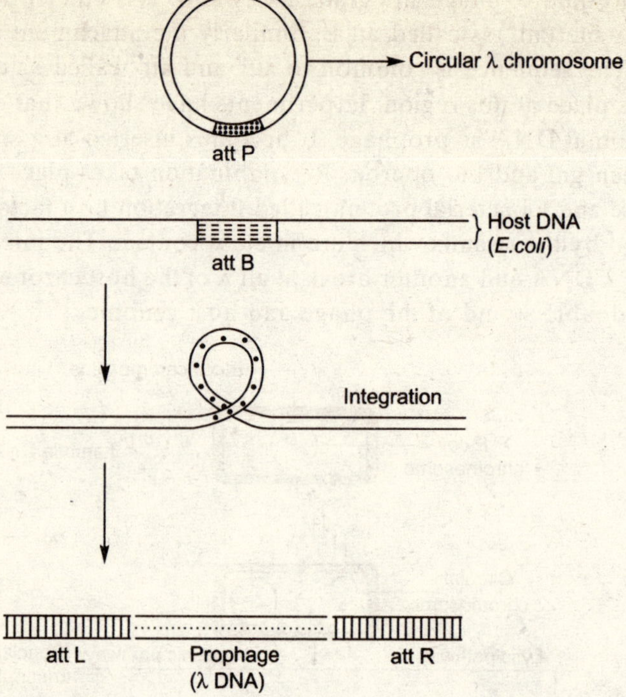

Fig. 5.3. *Site specific recombination during specialized transduction.*

During specialized transduction, recombination involves host chromosome and the phage chromosome, resulted in phage chromosome containing a piece of bacterial DNA. Thus, specialized transducing particle always contain both phage and bacterial DNA. The best example is λ phage mediate transduction of only the gal and bio genes of E. coli.

During excision this event occurs occasionally at a site other than the original attachment site. As a result a portion of the phage DNA is left on bacteria and portion of bacterial DNA is excised and is with the phage DNA. Due to this mistake specialised transducing particles are formed at packaging.

Amazing question is which genes of host bacterial DNA normally excised alongwith phage DNA. The answer is only host bacterial genes situated close to the site of prophage DNA insertion are excised. Thus specialized transduction involves only a transfer of bacterial genes situated close to integrated phage DNA (Fig. 5.4).

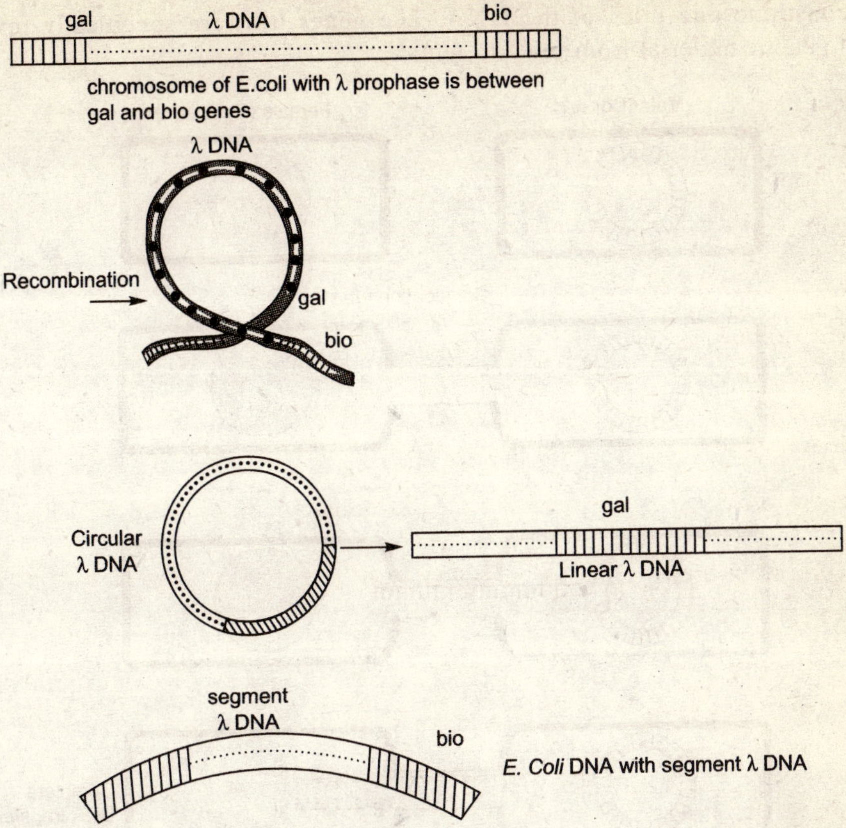

Fig. 5.4. *Specialised transducing particle.*

5.3 CONJUGATION

Conjugation was discovered by Lederberg and Tatum (1946). In this pòocess DNA is transferred from one bacterium (donor cell) to another bacterium (recipient cell) through specialized connection between the cells known as conjugation tube. For convenience the donor and recipient cells are also referred as male and female cells respectively i.e., the one which donate DNA is male and one receives DNA is female cell. It does not involve DNA exchange instead DNA is transferred from one to another cell in one way direction. During conjugation male and female cells come in close contact but do not fuse completely.

Distinguishing male cell from female cell can be observed by the presence of specialized cell surface appendages called 'F.Pili'. The cell which can donate DNA contains 'F.Pili' is male and the cell which receive DNA do not contain 'F.Pili' is female. Several genes are controlled in the synthesis of 'F.Pili' at least seven genes are carried by a small circular DNA or mini chromosome called 'F' factor i.e., fertility factor. Only cells carrying F factor form conjugation tube and initiate DNA transfer when close contact with female cells not carrying F factor. This female cell is also represented as F^-cells. The F^-factor is a closed circular double stranded DNA molecule present in one copy per cell. In F^+ DNA there are 19 genes known as tra-genes together constitute one third of the DNA. The genes (tra) are specifically involved in the transfer of genetic material from male to female.

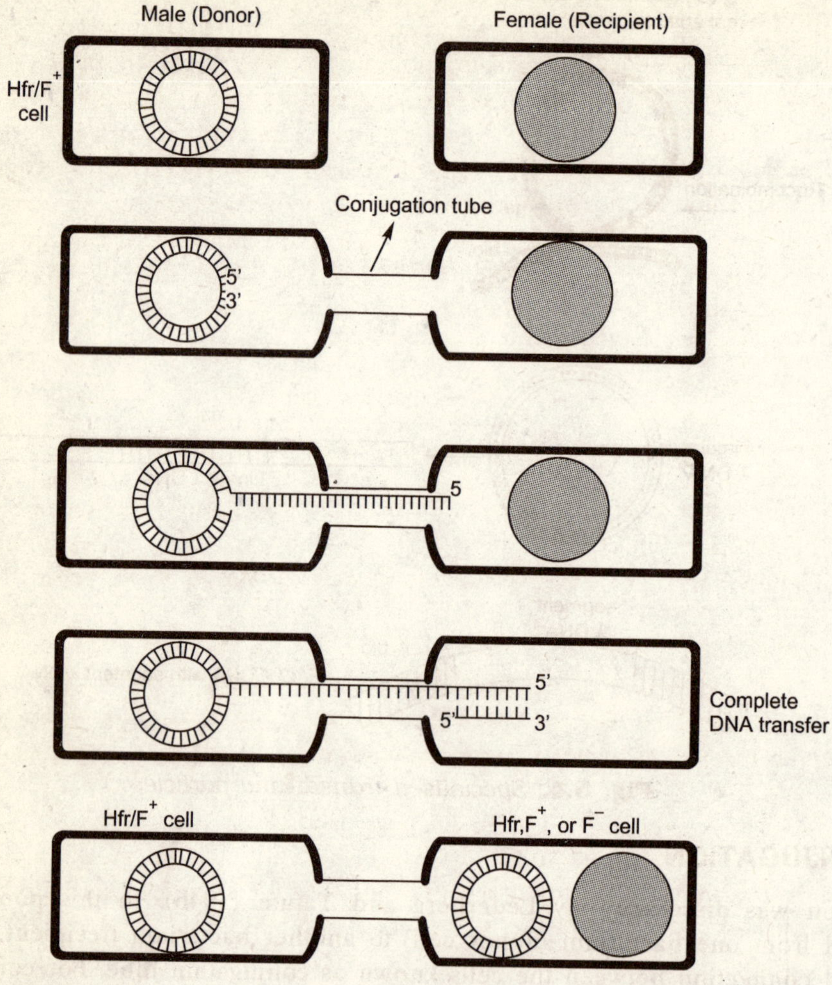

Fig. 5.5. *Mechanism of DNA transfer during conjugation.*

In the laboratory experiments when these two F^+ cells are mixed, conjugation proceeded by attachment of a male (F^+) sex pilus to the surface of F^- female cell. Attachment signals the F^+ bacterial chromosome to undergo replication and leads to the transfer of newly synthesised DNA of F^+ chromosome to the F^- cell. It is interesting to see that none of the genes present on the main chromosomes are transferred. Due to transfer of F factor to recipient F^-cells, also become male F^+ donor cells (Fig. 5.5).

5.3.1 Role of F' Factor in Gene transfer

The F factor can integrate into the host chromosome at any one of the site in site specific recombination event. The initiation of F factor is believed to be aided by IS elements. A cell carrying F factor is termed as Hfr i.e., High frequency recombination. The F factor mediates the transfer of a chromosome of the Hfr cell to a recipient (F^-) female cell.

5.3.2. Mechanism of DNA transfer

During conjugation the cell contact occur-it is followed by the formation of conjugation tube between two cells. An endonuclease cleaves one strand of DNA at a unique site on the F factor. The 5′ end of the cleaved strand is then displaced, 5′ end moves through the conjugation tube into the recipient cell. Transfer of DNA occurs concurrently with rolling circle replication. At this stage the intact circular strand serving as a template for the synthesis of a complementary in the donor cell and the transferred linear strand being replicated discontinuously in the recipient cell. In F^+ by F^- mating i.e., when F^+ F^- cells are mixed to allow conjugation both cells will be F^+ since both will have a complete copy of the F factor. In Hfr by F^- mating, the donor cell will remain Hfr and the recipient cell will contain F^-. Finally in the conjugation process entire chromosome is transferred.

5.4 ROLE OF REC A PROTEIN IN RECOMBINATION

Recombination in E. coli may takes place predominantly by breaking of DNA strand (nicking) and uptake of marked DNA. Detailed understanding of recombination comes from the studies of rec A gene product known as Rec A protein in E. Coli. This protein is capable of carries out some of the important steps in recombination. The Rec A protein specifically recognizes single stranded DNA. It causes uptake of single stranded DNA and anneal it to a complimentary sequence in a homologous regions of the duplex DNA. Rec A protein first binds to a single stranded DNA and the ATP is without being hydrolysed acts as a cofactor to this process. Single polypeptide of Rec A protein binds to five nucleotide of single stranded DNA, thus forming a DNA-protein filament (Fig. 5.6). This filamentous Rec A protein initiate branch migration. So that increasing single stranded DNA can displace one of the strand of original duplex helix. The presence of Rec A protein using hydrolysed ATP melt (unwind) double strand DNA at different region. More precisely when a single stranded molecule react with a duplex DNA, it becomes unwind in the region of recombinant joint.

Homology is found, annealing the bound single strand DNA to unwind regions. Once small portion is annealed, energy of the ATP is utilized in pairing between incoming single strand (SS) DNA with original helix, moving in 5′–3′ direction Rec A protein is then displaced and new hybrid DNA is formed (Fig. 5.6).

It has been found that Rec A protein stimulates branch migration. A free end is required for strand exchange and large amount of ATP is hydrolysed during this reaction. The presence of single strand protein (SSB) along with Rec A protein stimulate this reaction.

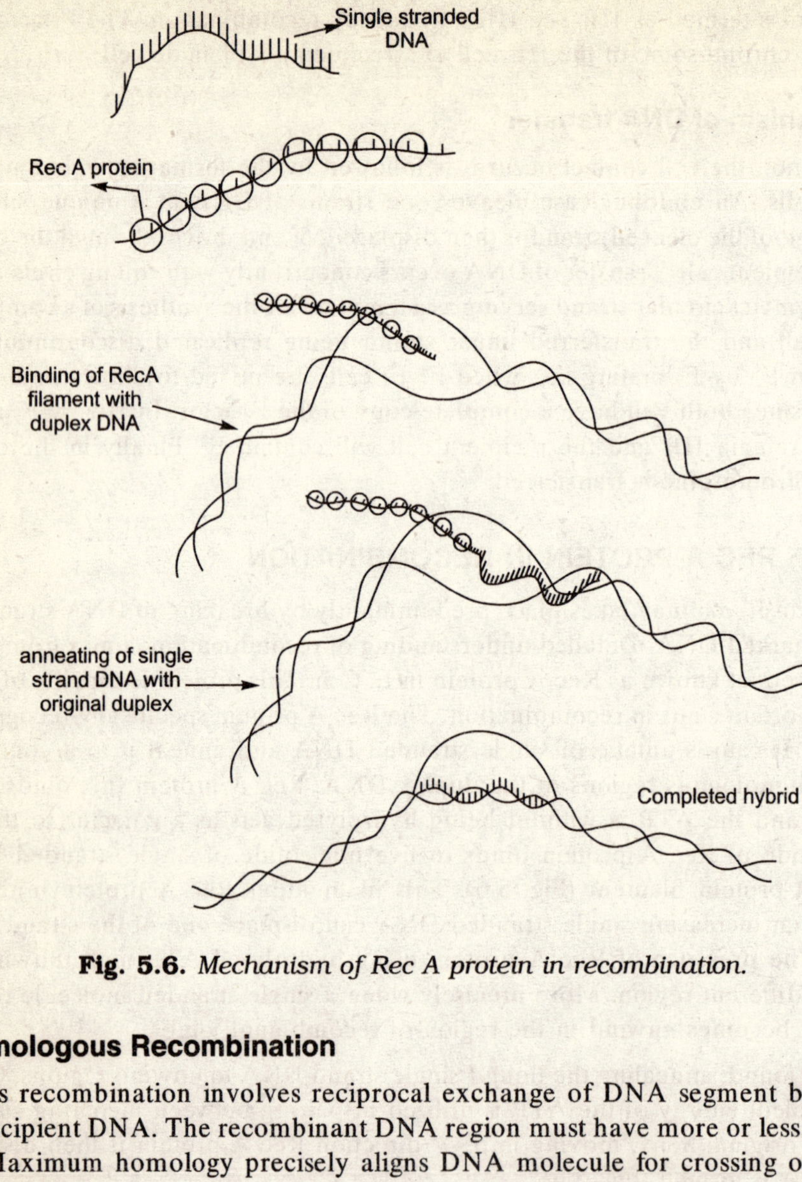

Fig. 5.6. *Mechanism of Rec A protein in recombination.*

5.4.1 Homologous Recombination

Homologous recombination involves reciprocal exchange of DNA segment between donor DNA and recipient DNA. The recombinant DNA region must have more or less the same base sequence. Maximum homology precisely aligns DNA molecule for crossing over. It creates genetic diversity by generating different combination of alleles (Fig. 5.7).

5.5 CROSSING OVER BY BREAKAGE AND REUNION (*HOLIDAY MODEL*)

Most of the popular model of crossing over are derived from the holiday model proposed in 1964. This also explaining recombination at molecular level. The mechanism involves breakage and reunion with repair synthesis. According to the holiday model crossing over begins with a endonuclease break in a single strands of each of the two parental DNA molecule. Segments of the single strands on one side are displaced from the complementary strands with the aid of DNA helix destabilization proteins and DNA unwinding proteins. The displaced strands are then exchange pairing regions of DNA. Exchanged strand are then base pairs with the complementary strands of the homologous chromosomes which is aided by certain protein known as Rec A proteins. In E. coli product of rec A gene referred as Rec A protein has been shown to facilitate reciprocal exchange of DNA strands between two DNA double helix. The Rec A protein mediates the process by binding to the unpaired strand of DNA and somehow search for homologous DNA sequence. Once a homologous double helix is found it triggers displacement of a segment of one strand of the double helix by the unpaired strand (Fig. 5.8). The studies here have shown that the Rec A protein will promote more than 50 fold increase in renaturation of DNA. The cleaved strands subsequently joined by DNA ligase.

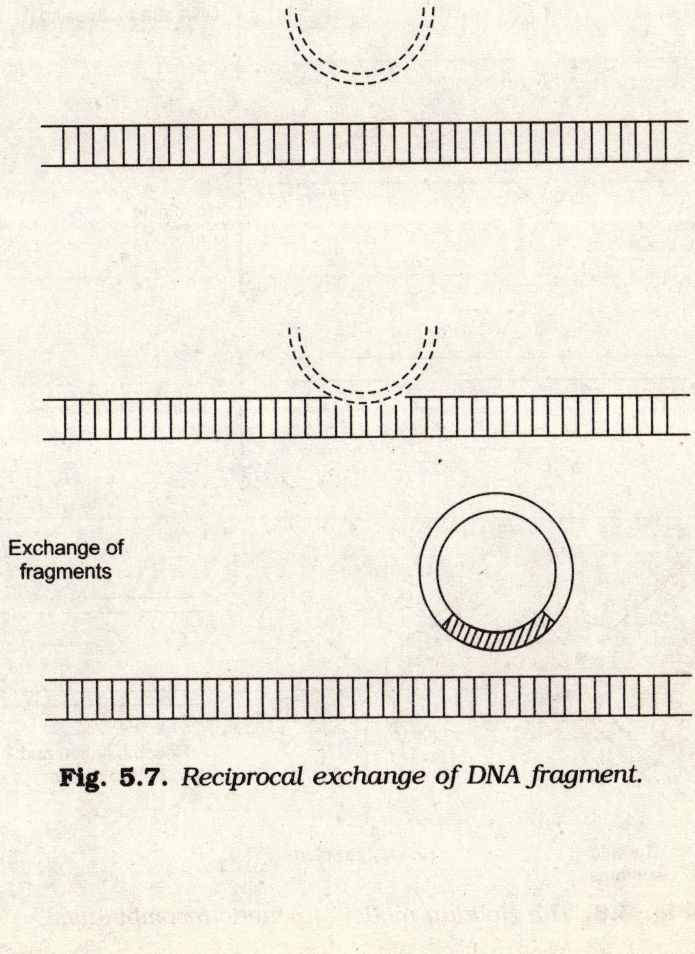

Fig. 5.7. *Reciprocal exchange of DNA fragment.*

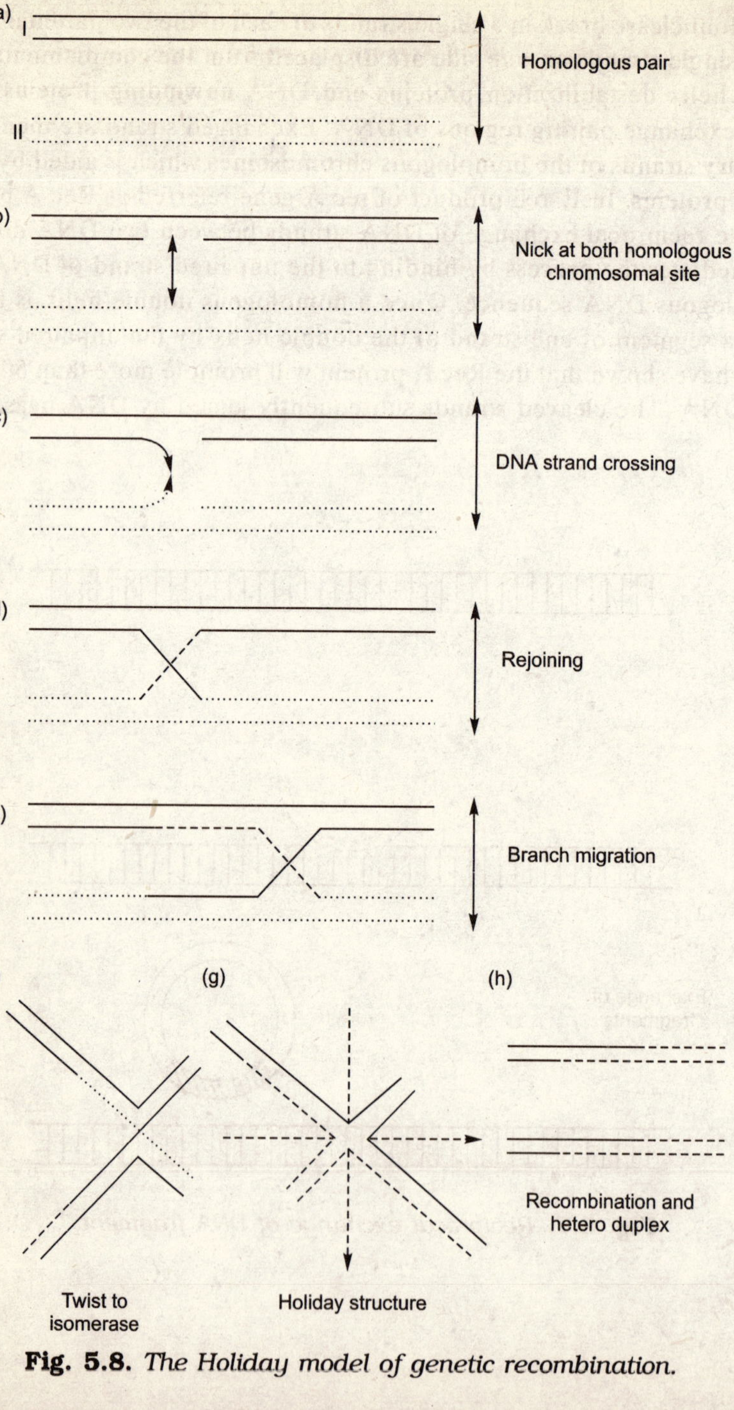

Fig. 5.8. *The Holiday model of genetic recombination.*

5.6 RECOMBINATION BY DOUBLE-STRAND BREAKAGE

The mechanism of recombination involve single strand break in the homologous pair is well

cells are the mitotic and probably meiotic phase, recombination proceeds with double strand break in the DNA molecules. The double strand break can initiate recombination in yeast even if the DNA studies have gap in them. In the pair of homologous chromosome I and II double strand break is created in homologous 1. The action of endonuclease produce gap in both the strand. It can produce 3′ protruding ends that can invade or enter chromosome II. Subsequently branch migration due to crossed strands leads to recombinations (Fig. 5.9). In the invitro experiment plasmids can be inserted with marker gene and is allowed to transform yeast cells. Such plasmid DNA is found recombined in to yeast genome at its homologous site on yeast chromosomes.

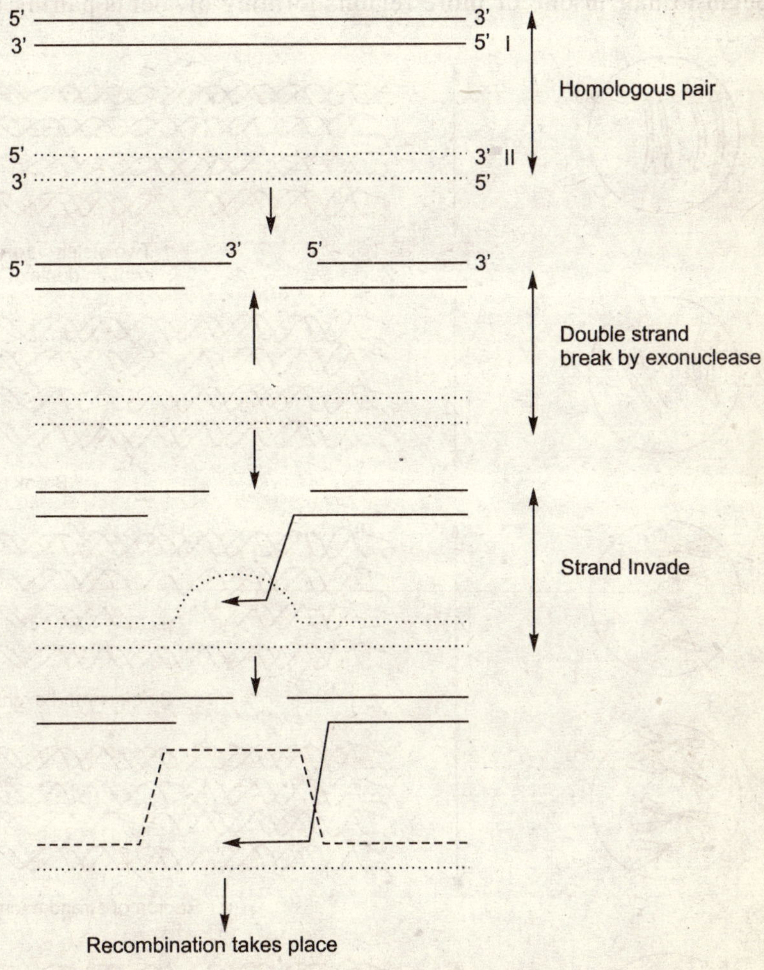

Fig. 5.9. *A double-strand break model for recombination.*

5.6.1 Role of Synaptemal Complex in Eukaryotes

A specialized structure in eukaryotic system is indispensable for recombination. The process of pairing between homologous chromosomes during meiosis may begin at similar points that become aligned when the chromosomes are close together at the nuclear periphery.

This phenomenon observed during meiosis known as synapsis, and this specialized structure called the synaptomal complex is visible in electron microscope. This complex is assembled in many eukaryotes to function as scaffold on which meiotic recombination might takes place.

5.7 MOLECULAR BASIS OF RECOMBINATION IN TYPICAL EUKARYOTES

To explain molecular basis events of recombination meiotic crossing over is an ideal event. Crossing over during first meiotic prophase set a stage for genetic recombination in eukaryotic system, leading to molecular interaction that are involved in exchanging genetic material between duplex of DNA. During meiotic phase the homologous chromosomes approach one another and begin to pair in one or more regions forming bivalents pairing take place in the

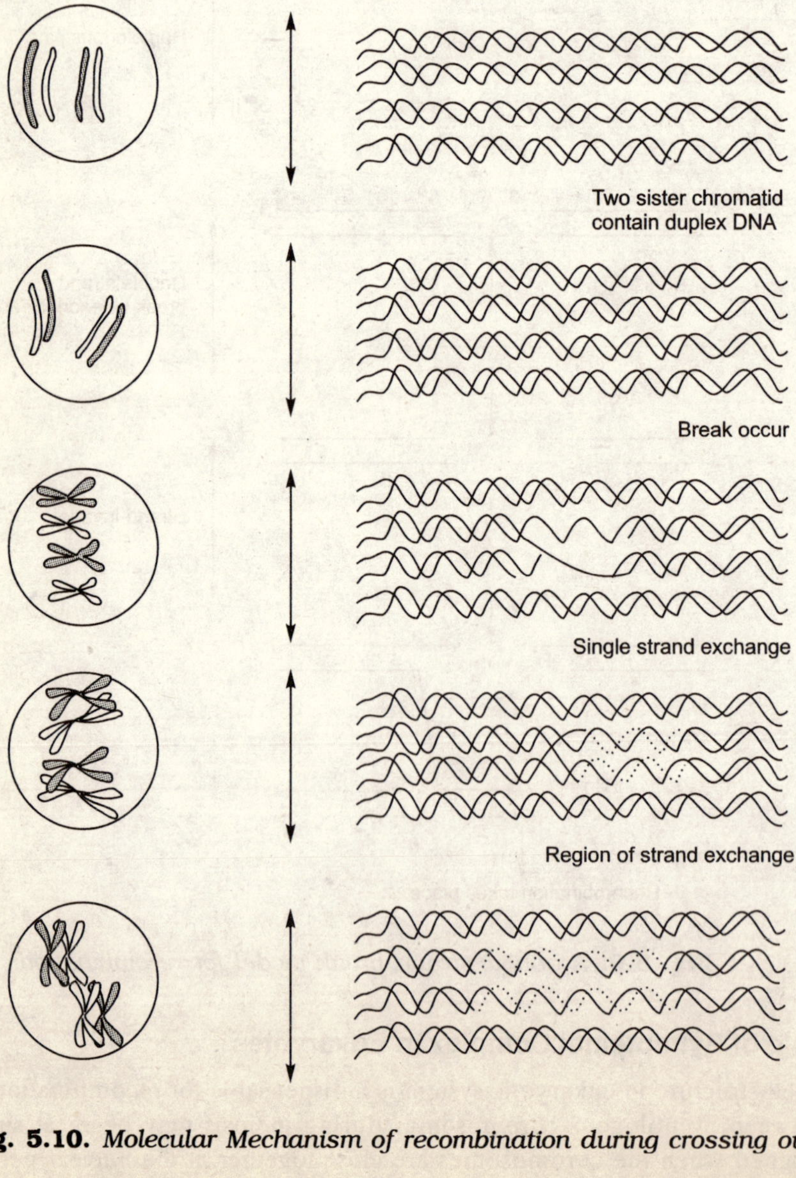

Fig. 5.10. *Molecular Mechanism of recombination during crossing over.*

entire length of the chromosome. The chromosomes are laterally associated in the form of synaptenemol complex (pachytene stage) at molecular level. The strand exchange begins at this stage when chromosome begin to separate. They held together at discrete sites known as chiasmata. The number of chiasmata indicates the feature of crossing over and region of exchange. When portion of DNA strand is exchanged genome is released by nicking.

To explain molecular basis of these event each sister chromatid contains a single double helical DNA (duplex). Thus each bivalent contains four double helical DNA. Recombination at molecular level can be seen when duplex DNA of one sister chromatid interact with duplex DNA of sister chromatid of other chromosome. First step is the break in each duplex DNA then one or both strand of duplex can be released. At least one strand of DNA displaces the corresponding strand in other duplex DNA. This would facilitate single stranded exchange. By exchanging both strand, it is possible to convert the parental double strand DNA molecule by cross over which involves breakage and reunion (Fig. 5.10).

SUMMARY

Recombination takes place in all living organisms. Recombination aids in speciation during the course of evolution. This process takes place even in cell organelles. Genetic recombination takes place in bacteria due to three types. They are transduction, transformation and conjugation. Transformation is the uptake of alien nacked DNA by bacteria with thc aid of enzymatic active process. It is physical inserted into host DNA. Integration results in the replacement of one of the strand of recipient. Transduction involves integration of viral λ DNA into the bacterial chromosome. It may show two pathway during integration. One is lytic and another is lysogenic pathway. This recombination involves breakage and reunion of chromosomal DNA. Integration of viral DNA takes place at specific attachment site. During conjugation F factor play a important role is gene transfer. As a result of conjugation when H Fr chromosome is transfer to recipient female cells and it becomes male F^+ donor cells.

Recombination process is predominantly aided by RecA protein. It can binds to single strand DNA and attaches to their homologous sequence and completes the hybrid. The crossing over mechanism based on holiday model involve breakage and reunion with repair mechanism. In addition due to endonucleolytic cleavage segments of the single strand on one side are displaced from the complementary strands with the aid of proteins and enzymes.

Eukaryotic crossing over requires synaptemal complex. Recombination takes place by double strands breakage and reunion. The strand exchange takes place during meiotic prophage stage.

QUESTIONS

1. Explain the molecular mechanism of recombination in prokaryotes.
2. Describe the process recombination in eukaryotic system.
3. Explain holiday model for crossing over during recombination.
4. Explain the molecular mechanism of recombination. (BU April 2000)

SHORT QUESTIONS

1. Holiday model
2. Rec A protein
3. Crossing over
4. Transduction
5. Transformation
6. Site specific recombination

6
CHAPTER

Structure of Prokaryotic and Eukaryotic Gene, Genetic-Code

Until the birth of molecular biology, it was strongly believed that chromosome contains linear arrays of genes. It was presumed that some non genetic material acts as binding force to hold these genes. However, realities on the exact nature of genes on the chromosome stems from the classical experiments of genetic crosses and the advent of recombination DNA procedure. After the discovery of double helix in 1950, several ambiguity associated with the nature of genes was cleared. Structure of the gene can be describes at the back drop of crossing over between the genes and also within gene. In 1909 Danish geneticit studied the assortment of Mendelian units of heredity. In the same line of approach, Thomas Morgan and his associate threw light on some of the basic nature of chromosomes and characters of genes. According to Morgan, one particular chromosome was responsible for sex determination and discovered correlation between number of linkage groups to number of chromosomes. For example, in Drosophila, four linkage groups, represent four visible chromosomes. Similarly in 1931, Barbara Macclintok discovered visible arrangement of gene segment within and between chromosomes with the distribution of specific traits in Zeamays. After this discovery and contribution from the other biologist, it has been established that genes are not hypothetical or theoretical entities but a well accepted biological entities. We now know that chromosome contain, single DNA molecule carrying information for the synthesis of proteins by specific arrangement of amino acids. However, genetic information can be provided by DNA or RNA. Cells use only DNA whereas viruses use RNA and its replication takes place in host cell.

6.1 SALIENT FEATURES OF GENES

Based on the elucidation of structure of genes following conclusions were drawn.

(a) Basically genes are DNA. The genetic nature of DNA was proved in 1944 by Avery and his associates. They proposed several pieces of evidence about chemical nature of genes.

(b) Genes are made up of DNA but never from protein. It can only direct to order synthesis of protein using its own message. The DNA present in all cell types in an organism is constant.

(c) Gene is a unit of recombination. The gene was considered as a shortest segment of chromosome. The crossing over can take place between the genes and also within a single gene. One of the subunit of gene is referred as recon. More precisely recon can be interpreted as smallest unit capable of recombining genetically. Recon may consist of one or two pairs of nucleotide. In 1955, Benzer proposed strong evidence that crossing over could take place within a single gene against earlier concept on crossing over only between genes. Based on crossing experiments in T_4 bacteriophage DNA, Benzer proposed that crossing over can take place at the lowest frequency range at 0.02%. Such a lowest percentage could be attributed to crossing over between just two adjacent nucleotides.

(d) Gene is a unit of function, otherwise known as cistron. According to Beadle and Tatum's gene is a functional unit for coding the synthesis of single enzyme. This one gene-one enzyme hypothesis was well received and expressed in diversified way. Since genes are expressed and synthesize messenger RNA together, it becomes one gene one messenger RNA. When several genes produce only one mRNA as single unit they are known as polycistronic mRNA. When protein or enzyme consist, of more than one subunits then one gene one enzyme or one gene one protein hypothesis could not be justified. This hypothesis becomes presently modified as one gene one polypeptide. It was also noticed that expression of some of the genes produce ribosomal RNA and transport-RNA instead of mRNA. Thus further modification has been made to propose that one gene one product is widely accepted.

(e) Gene is a unit of mutation otherwise known as muton. Alteration or change in nucleotide sequence is possible in DNA. The sequence of DNA when undergoes mutation (muton) then alteration or change in nucleotide sequence is possible in DNA. The sequence of DNA which undergoes mutation has been called the muton, even smallest number of nucleotides (or single nucleotide) can undergo change. This is referred as point mutation (muton).

(f) Product of expressed gene is not always a protein. Instead, it may produce an rRNA or tRNA. Hence, this statement was further modified into one gene, one product.

(g) Gene can undergo duplication during its replication. Gene duplication takes place during interphase after mitosis. During replication duplex helix are separated due to unwinding factors. A new complementary strand is synthesised.

(h) By comparing gene and protein, gene is not directly translated into protein, but is expressed via production of messenger RNA. It is the nucleic acid intermediate used to synthesise protein after undergoing translation. The sequence of messenger RNA is complementary to the sequence of one strand of DNA and is identical with other strand of DNA.

(i) Genes can be mapped by restriction cleavage. A restriction map represents a linear sequence of the site at which restriction enzyme cut at the specific target sites. The distance between such maps is measured directly in base pairs (bp).

(j) As we know that a change in base pair in a gene can lead to mutation and this gene is known as mutant gene. Most mutant genes are recessive with respect to wild type genes. This can be evidenced from recessive phenotype (absence of particular phenotypic character). Due to failure of mutant gene to produce any functional proteins or enzymes.

However, in heterozygotes, there is often number of good genes and number of good proteins.

(k) In the back drop of classical genetics concept of dominant gene and recessive genes have been well established. For example, in heterozygote i.e., the presence of two genes one from paternal and other from maternal side. Then one good number of genes, able to express readily by suppressing the expression of recessive genes. It is possible that there is always a good copies of a relevant proteins in heterozygotes than in individuals with two wild types genes.

(l) Eukaryotic genes are interruptive i.e., coding sequence are interrupted by the presence of non-coding intervening sequence are known as introns. These intervening sequence will be eliminated within the nuclear region to produce mature RNA.

(m) The interrupted genes are otherwise known as split genes. Many genes are more introns than exon. The tRNA and rRNA genes also contains introns.

6.2 FINE STRUCTURE OF PROKARYOTIC GENE

Structure of viral gene : Considering elucidation of fine structure of gene, the virus (T_4 phage) and E. Coli is being considered as better and reliable organisms among prokaryotes. A simple phage contains atleast few proteins coding genes. Viruses can synthesise mRNA within host cell example. Some of the bacteriophage like k17 contain RNA as genetic material. This is single stranded RNA contains only three genes. These three genes (1.2.3) are being expressed in host cell and produce a protein (gene one), coat protein (from gene two) and subunit of replicase (gene three) (Fig. 6.1). Similarly polyomaviral chromosome measure 1-7 micron having 5-10 genes distributed on their chromosome. Another single stranded DNA $\phi \times 150$ virus contains genes codes for a proteins. However, gene number was found to be slightly increased among viruses example in bacterial virus lambda phage (λ) has about 11 gene and T_4 viruses contains more than 150 genes. Least number of gene i.e., only one is present in single stranded RNA virus STNV containing 1,300 base pair. Among single stranded RNA virus influenzae contain 12 genes with 13,500 base pairs. Greatest number of genes can be found in double stranded DNA virus i.e., vaccinia virus having less than 300 genes with 180,000 base pairs. Among double stranded RNA virus Reovirus possess 22 genes with 23,000 base pairs. Viroids are infections agent that cause disease in higher plants contains small circular RNA. This acts as a infections agent with 300 base pairs.

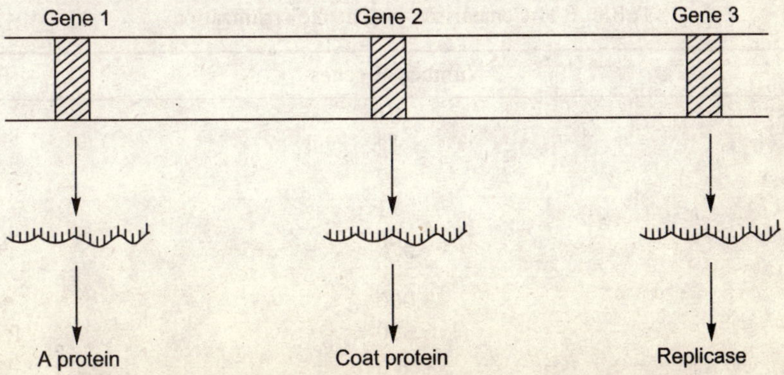

Fig. 6.1. *Simple gene organization in certain phage.*

6.2.1. Structural Organisation of Bacterial Gene

Organisation of genome interms of increase in number of genes have taken place from viruses to bacteria during evolution of life. The gene number among bacteria varies between 2000-4000 with 10^7 base pairs. In Escherichia coli for example contain single chromosomes is about 1mm long and comprises about 4000 genes with 4,000,000 base pairs or < 10^7 base pair.

Bacterial genes have organized as operon where genes associated with pathways are clustered on DNA i.e., several genes are present adjacently in a particular region more or less tightly associated, but not overlapped. These are known as structural genes or operon. The operon site contains a crucial control sites. They may be under repressed or induced state.

The control sites are promoters and operator. In addition, controlled sites are regulated by the presence of regulatory gene situated close to control site. In bacteria there is a signal for RNA synthesis and also signal for its termination. In general structural organization consist of regulatory genes, control site and structural genes. These genes are expressed in highly coordination fashion depending on their surrounding signal.

The action of bacterial genes can be noticed during synthesis of mRNA molecule. There are specific signal that start RNA synthesis are called promoters. Their existence was first revealed by the discovery of mutants blocked in the initiation of transcription. Bacterial promoters are positioned just before the starting site for transcription. The gap between gene and promoter is perhaps about - 10 base pair and - 35 base pair in the upstream direction. Due to this gap promoter sequence are transcribed in mRNA. There are special regulatory sequences often situated between promoter and amino acid coding sequences are called operator, to which specific DNA binding protein known as repressor can bind. The presence of repressor on this region can stop binding of RNA polymerase. Regulatory gene sequence always keep the operon in repressed state by producing repressor protein shows affinity to bind to operator sequence. The signal sequence used to stop RNA synthesis is highly complicated. This stop signal are located before the actual termination sites.

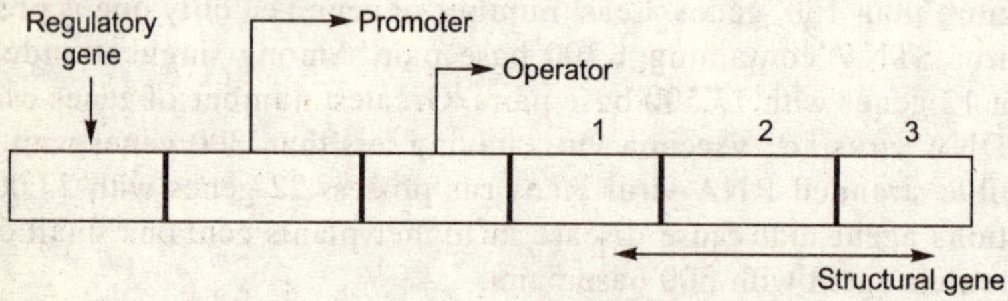

Table 6.1 Comparison of genomic organization

	Genome	Number of genes	Base pairs
1	TMV virus	01	6,500
2	Influenzae virus	12	13,400
3	Reovirus	22	23,000
4	Phage $\phi \times 174$	11	5,380
5	Phage T_4	200	165,000
6	Mycoplasma	500	< 106
7	Bacteria	3,000	
8	Fungi	6,000	1.3×10^7
9	Worms	14,000	$- 10^6$
10	Mammals	1,00,000	$- 3 \times 10^8$
11	Plants	50,000	$- < 10^{11}$.

6.2.2 Gap between Repressor and Genes

Usually there is low frequency of crossing over between mutation located at adjacent ends of two genes within a bacterial transcriptional unit or operon. This finding drawn to the conclusion that adjacent mgenes within operon are often separated by only a few base pairs. For example in tryptophan operon the gap between adjacent genes and several nucleotide are called as spacer DNA, would carry no function and would increase the time reqired for replication of bacterial chromosome.

6.3 FINE STRUCTURE AND ORGANIZATION OF EUKARYOTIC GENOME

First of all there is a considerable increase in the amount of DNA in eukaryotes. Eukaryotic cell can access to vast amount of DNA the amount present in per higher eukaryotic cell is almost thousand times greater than E. coli. Hence, vast amount of DNA is access to greater degree of organization and complexity of genome. But there is no reason to believe that any increase or decrease in gene size, substantial increase in vast number of different genes are perhaps the main contribution to complexity of genome.

Eukaryotic cell can initiate mammoth task of producing million's of proteins. They are of course capable of performing this job. Eukaryotic cells are distinguished from those of prokaryotes by the presence of nucleus, separated from the rest of the cell by a clear cut boundary known as nuclear membrane. The DNA is organized in linear chromosome rather than in a single circular chromosome. The size and number of chromosome differ from one species to another species. The total content of DNA of an organism in evolutionary position is redundant, since some fish, amphibia having a possession of larger genome than humans and other mammals.

The chromosomes consist of two complementary anti-parallel strands of DNA and are associated with good amount of proteins. Chromosomes exist in pairs while male possess an X and a Y chromosome, differing from each other by their size and structure. Haploid cell containing only half the genetic material or somatic cells which are in diploid status.

6.3.1 Eukaryotic Genes are Interrupted (*Split Genes*)

Bacterial genes and proteins are collinear. By comparing the nucleotide sequence of a gene with the amino acid sequence of a protein. We can draw a conclusion that sequence of a nucleotide in a gene corresponds exactly with the sequence of amino acids in protein. These are of course a statement with reference to prokaryotes. On the contrary when this colinearity concept is checked among eukaryotic some discrepancies were found i.e., when comparison between the structure of DNA and the corresponding mRNA was assessed. It shows marginal difference. This is slight deviation from the rules of the genetic code. After careful investigation it shows that genes includes additional sequences that lie within the coding region. The sequence of DNA comprises interruptive or intervening sequence are divided into two categories. They are exons, and exons are true sequences represented or can be seen in mature RNA. Exons can also defined as, a gene starts and ends with exons. Another category is the introns. These are intervening sequence that are removed when primary transcript is processed to give the mature RNA. Since, introns does not code for any protein. They must be removed in mRNA itself before translation. Intervening sequence of intron on DNA will be transcribed into its corresponding mRNA. The fate of introns are then decided by spliceosome factors.

Once introns are eliminated by RNA splicing, exons are always joined together in the same order in which they organised in DNA. So after RNA maturation the colinearity of gene and protein is maintained between the exons and corresponding proteins.

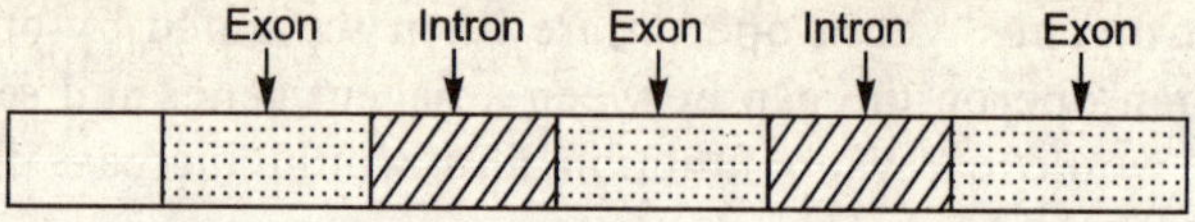

All types of genes contain intervening sequences i.e., introns. These interrupted genes may codes for proteins, ribosomal RNA and transport RNA. Introns can be seen even in organelle DNA like mitochondria of lower eukaryotes. Interrupted genes are extremely rare in prokaryotic genome. It is a matter of debate on how many introns are present in a gene. It may be between one to several introns for example in globin gene after RNA splicing two introns get eliminated followed by joining of three exons. It may be seen that in an extreme cases. Exons of two genes may code for the same gene but their introns may be different.

Number of interrupted genes have been found to be increased drastically from lower eukaryotes like yeast to mammalian system. The percentage of exons in yeast attains more than 95% when compare to the mammalian system, where its share is less than 4%. More than 50% of mammalian gene contains more than ten introns. It was a matter of speculation that how did interrupted gene are evolved. Couple of models have been proposed do strengthen the idea on its origin. According one model introns are integral part of the gene. So that they were exist in ancestral genes. However, another model claims that ancestral protein coding gene consist of uninterrupted sequence and introns were inserted during the course of evolution.

6.3.2. Size of Eukaryotic Genes

Considering the genetic status of haploid. The total amount of its genome is a characteristics of each living species known as C-value. The C-value paradox denotes the discrepancy between genomic size and genomic complexity. There is a considerable degree of variation of C-value from prokaryotes mycoplasma (< 10^6 bp) to eukaryotic amphibians (> 1011 bp). Substantial increase in gene numbers could be sighted between simple mycoplamsa (500) to mammals (100,000). Plants also exhibit substantial possession of around or less than 50,000 (< 1011 bp). The intermediate life forms like worms containing nearly 14,000 genes (- 10^8 bp). Investigation on human genome map reveals that human DNA from 46 chromosomes claims to have contains 30,000 to 40,000 genes as against previous prediction of 100,000 genes (Table 6.1). The size of the genome can be estimated by denaturing the DNA and allowing it to reanneal. The larger the size of genome, the longer it will take homologous sequence to reanneal.

6.3.3 Eukaryotic DNA Contain Repetitive Sequence

Based on denaturing and reannealing of DNA, careful analysis reveals that eukaryotic DNA contain repetition sequence from moderate to highly repetitive state. For example in mouse ten percent of short nucleotide sequences were repeated in million copies. Based on repetitive status of DNA at least three classes of DNA have been categorized as follows.

Highly repetitive DNA : These are short sequences of DNA generally found in satellite DNA. In eukaryotes, satellite DNA are present in thousands to million copies per cell. Satellite

DNA can be separated from the rest by caesium chloride centrifugation. It is due to their base composition difference as a reflection of buoyant densities. Position of satellite DNA in a chromosome is in heterochromatin region nearer to centromere position. The number and size of satellite DNA can correlated to smaller or larger heterochromatin region. Satellite DNA are formed by sort nucleotide sequence that are tandomly repeated in DNA.

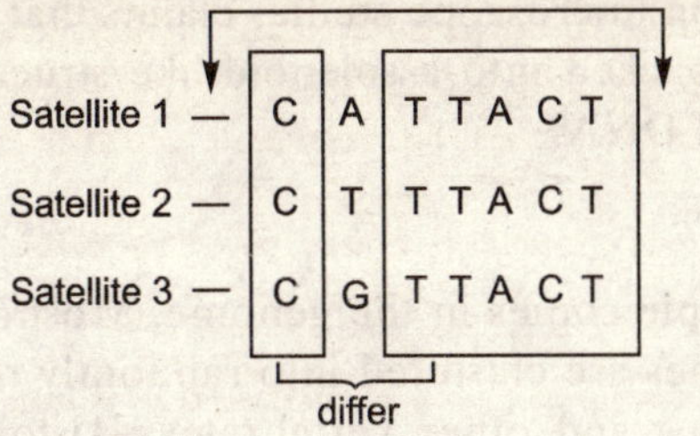

One of the species of Drosophila has 12 million copies per haploid genome, of the 7 nucleotide repeat of satellite I i.e., ACAACT. During transcription most of the sequence of satellite DNA are not transcribed into RNA. Their exact function is not known but they tend to undergo reshuffle in their nucleotide sequence. Today in modern biology these satellite DNA and more precisely minisatellite DNA acts as powerful tool in identifying and finger printing of individual organism and genetic diversity.

6.3.4. Moderately Repetitive DNA

Some of the genes are present in multiple copies i.e., several hundred times. Some of the amphibians contain 450 rRNA genes and 20,000 5S RNA genes are associated in clustered form. These are separated by AT rich spacer DNA. Most protein coding genes are present only once per haploid genome. However, gene codes for the 5 histone proteins are repeated about 400 times in sea urchin. These histone genes repeated between 10-20 times in human DNA.

In general, ribosomal DNA, 5S DNA and histone gene contain moderately repetitive DNA. Transpons are also the main component of repetitive DNA. They have either direct or invert repeats. In Drosophila majority of the moderately repetitive DNA is formed by 10-13 families of transposable elements.

In human the most important and abundant mobile element is Alu sequence of about 2,50,000 copies of sequence of 300 nucleotide. This Alu sequence is flanked by direct nucleotide repeats.

6.3.5. Single-copy gene

These are mostly protein coding genes and their introns. These single copy DNA accountable upto 60 to 70% of the mammalian genome. Cells can make a new copy of genes by a process known exon shuffling. In this process two exons or two existing genes are recombined to generating new gene. In general protein coding genes do not contain repetitive sequence.

6.3.6. Eukaryotic DNA is Associated with Proteins

The nuclear DNA of eukaryotes is associated with a number of proteins. These proteins are generally found within the nucleus. One of the most well identified proteins present within the nucleus is histone. There are five major classes of histone proteins containing excess positively charged ones due their lysine and arginine amino acid. Excess positively charged histone proteins binds with negatively charged nucleic acid by multiple ionic interaction.

The DNA structure in the nucleus appears as nucleosome and looks like series of beads on a string. The beads are 9-11 nm in diameter. The DNA in these fashion follows the course one and three quarter turns of a left-handed helix wrapped round the outside of an octamer consisting of two molecules each of histones (H_2A, H_2B, H_3 and H_4). Histone H_1 is bound to the outside of nucleosomes and probably play a role in bringing individual nucleosomes together for stabilization. Electron microscope studies claims that they form a fibre of 25-30 nm in diameter in which they are packed into a solenoid-like structure. These fibres are formed into domains of size 35-85 kb of DNA.

6.3.7. Histone Genes

They are usually found in multiple copies in the genome. Most of the experiments in animals study shows that the histone genes are clustered into randomly repeated units. However, they are moderately repeated in yeast and other vertebrates. Histone genes contain no introns. Their mRNA are not polyadynylated.

6.3.8. tRNA Genes

tRNA genes are also present in multiple copies in both nuclear and organellar DNA. In xenopus many of these occur at about - sites each of about 3200 bp long and containing 8tRNA in tandom repeats. Some tRNA genes are in cluster or singly.

6.3.9 Hormone Genes

The genes for several hormones have been well characterized in higher eukaryotes. Some of the classical example for hormonal gene are-Insulin, human growth hormone, thyroid stimulating hormone and follicle stimulating hormone etc.

Human insulin is encoded by a single gene which has two introns one in the 5′-flanking region and a long one in the coding region. This intron is longer than the coding part of the gene. The 5′-flanking region sequence of the gene is highly polymorphic. The rat has one gene for growth hormone and one for prolactin. They contain five exons and four introns. Two forms of growth hormone (GH) are found in human pituitaries which differ in that one has a deletion of aminoacid residue 5′ 32-46. This sequence is encoded at the beginning of the third exon.

6.4 STRUCTURE OF PLANT GENOME

Organisation of plant genome is like that of any other eukaryotes. As in other systems plant genes are described under two headings. One is protein coding genes and other is RNA-coding genes. In plant genome, transcription can start 40-80 nucleotide upstream of the ATG initiation codon. Plant genes also carries no-coding sequence or introns. These are removed successfully at the post-transcriptional level to form mature of mRNA. The genes have controlling sequences at the DNA at the 5′-3′ ends. The 5′ region has many regulatory elements and are unique for specific genes. These are otherwise called as core sequences of promoters. These two regions are commonly found in all plant genes are called TATA and CAAT boxes. The TATA box is situated 25-40 nucleotides upstream of the transcription initiation site. The consensus sequences of the TATA box is TCACTATATATAG. The CAAT box is located further upstream from the start of transcription. In some of the plant genes instead of CAAT box it is replaced by AGGA box. The coding region of the mRNA begins with an AUG codon which is part of the conserved sequence AACAAUGGCT. The 40S ribosomal subunit can scan the mRNA from 5′ direction and select this conserved sequence for initiation of translation (Fig. 6.2).

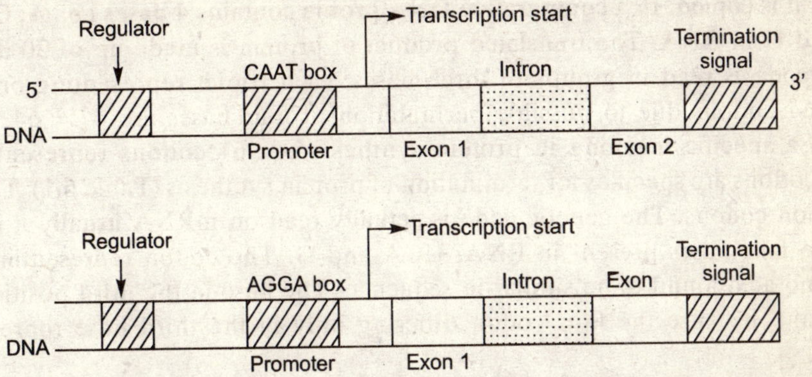

Fig. 6.2. *Structural organization of plant gene.*

6.5 GENETIC CODE

Deoxyribonucleic acid (DNA) is the genetic material of all the cells. However, exceptions are RNA viruses. It carries information in coded form, passes from one generation to another i.e., from parent to offsprings. The sequence of DNA strand read from 5′-3′ direction consist of nucleotide triphosphate corresponding to the amino acid sequence of a protein read from N-terminus to C-terminus. While coded sequence in a DNA when expressed, its information is passed first into another nucleic acid known as ribonucleic acid (RNA). The information present in RNA is then translated to produce proteins. It is the ultimate gene product. The passage of information from DNA to RNA and then to proteins as final products is known as "Central dogma of molecular biology. Although DNA is the genetic material of cells in general, certain viruses possess RNA as the genetic material.

$$\text{DNA} \xrightleftharpoons{\text{Transcription}} \text{RNA} \xrightarrow{\text{Translation}} \text{Protein}$$

Genetic code can be stated as the relationship between sequence of DNA and the sequence of corresponding protein. Codons are carried on mRNA and are therefore also found on the DNA strand that is not transcribed. It is possible to compare codon sequence on DNA with amino acid sequence on protein directly.

Genetic code is universal hence, coded message is the same for all living organism. Deciphering of genetic code is one of the most remarkable discovery in biological field. Breaking of genetic code shows that nucleic acid is a information molecule and genetic information is stored in the form of nucleotide triplet, known as triplet codon. Persistent effort was made by Nirenberg in 1966 due to wide range of biochemical experiment to define the term codon and considers how the four-letter language of nucleic acid is translated into twenty letter language protein.

6.5.1 CODONS

Codon is the hereditary units that contain information coding for one amino acid. Each codon consist of three nucleotide (a triplet). This codon information on DNA is first transcribed into messenger RNA. The sequence of bases in mRNA are complementary to the bases in DNA from which it is copied. In a comparative DNA profile contains 4 bases i.e., A, G, C and T and A, G, C and U in RNA. The translated product of protein is made up of 20 different amino acids. The code is read in groups of three bases, each triplet representing one amino acid. There are 64 codons due to possible permutation of four bases are $4^3 = 64$. Each of these codons has a specific meaning in protein synthesis i.e., 61 codons represent amino acids. Whereas 3 codons are specifies for termination of protein synthesis (Table 6.1). They are known as termination codons. The genetic code is actually read on mRNA usually it is described in terms of the four bases present in RNA, UCA and G. The codon representing the same or related amino acid found to be similar in sequence. The base in the third position of a code is not significant, because the four codon differing only in the third base represent the same amino acid.

The length of codon portion of the gene depends on the length of message to be translated i.e., number of amino acids in the protein. For example DNA contain gene sequence of 1500 nucleotide may contain 500 codons. (Three nucleotide forms one codon), that code for a protein having 500 amino acids. (one codon for one amino acid). Therefore the sequence of triplets determining sequence of amino acids in a protein. The explanation of a codon requires base pairing with anticodon of the aminoacetyl-tRNA.

Table 6.2 Genetic code

	U	C	A	G	
U	UUU - Phen. UUC - Phen. UUA - Leu. UUG - Leu.	UCU - Ser UCC - Ser UCA - Ser UCG - Ser	UAU - Tyr. UAC - Tyr. UAA - Oche UAG - Amber	UGU - Cys UGC - Cys UGA - Nonsense UGG - Trip	U C A G
C	CUU - Leu. CUC - Leu. CUA - Leu. CUG - Leu.	CCU - Pro CCC - Pro CCA - Pro CCG - Pro	CAU - His CAC - His CAA - Glu CAG - Glu	CGU - Arg. CGC - Arg. CGA - Arg. CGG - Arg.	U C A G
A	AUU - Ile AUC - Ile AUA - Ile AUG - Met	ACU - Thr ACC - Thr ACA - Thr ACG - Thr	AAU - AspN AAC - AspN AAA - Lys. AAG - Lys.	AGU - Ser AGC - Ser AGA - Arg AGG - Arg	U C A G
G	GUU - Val GUC - Val GUA - Val GUG - Val	GCU - Ala GCC - Ala GCA - Ala GCG - Ala	GAU - Asp GAC - Asp GAA - Glu GAG - Glu	GGU - Gly GGC - Gly GGA - Gly GGG - Gly	U C A G

Genetic code is commaless i.e., it does not contain any punctuation or signal required to indicate the end of one codon and the beginning of the next. Only initiation signal for RNA synthesis and termination signal to stop RNA synthesis are present in the form of codon.

6.5.2 GENETIC CODE IS Degenerate

In the universal genetic code system many amino acids are specified by more than one codon, a phenomenon called degeneracy for example, two triplets codon UUU and UUC code for phenylalamine, while serine amino acid is coded by six triplet codon namely UCU, UCC, UCA, UCG, AGU and AGC when the first two nucleotides are identical. The third nucleotide can be either cytosine or uracil and the codon will still code for the same amino acid (Table 6.3). However, for some amino acid degeneracy of genetic code based on equivalence of the first two nucleotide is not applicable. Leucine for example is a codon by UUA and UUG, as well as by CUU, CUC, CUA and CUG. Codon deneracy explains variations in AT / GC ratio in the DNA of several organism without much changes in the proportion of amino acids in their proteins. In some codon for amino acids third position is filled by purine of third position of other by pyrimidine forer. The codon for histidine purine of third position of other by pyrimidine forer. The codon for histidine is CAU and CAC. The third bases are pyrimidine i.e., U and C and for glutamine codon is CAA and CAG. Third position is occupied by purines A and G.

6.5.3 WOBBLE HYPOTHESIS AND SIGNIFICANCE OF WOBBLE IN THE tRNA

Earlier studies proposed that a specific tRNA anticodon would exist for every codon. Suppose of that were the situation, all the cells would have equipped with at least 61 different tRNA, and additional three for termination. Later studies confirmed that tRNA species could recognize several different codons. It was also discovered that one of the anticodon base was different and it was not like four regular ones. This fifth base is known as inosine arises through enzymatic modification of a base present in an completed tRNA chain. It is derived from the adenine base. In order to explain these phenomenon wobble concepts was proposed in 1966. According to Wobble concept base at 5′ end of the anticodon form hydrogen bonds with any of several bases located at the 3′ end of codon. Only certain pairing is possible. For example U at the wobble position can pair with either adenine or guanine. The inosine (I) can pair with U, C or A. The Wobble rules do not permit any single tRNA molecules to recognize four different codons. Three codons can be recognized only when inosine occupies the first position of the anticodon.

Further evidence gathered that at least three tRNA's exist for six serine code words like UCU, UCC, UCA, UCG, AGU and AGC.

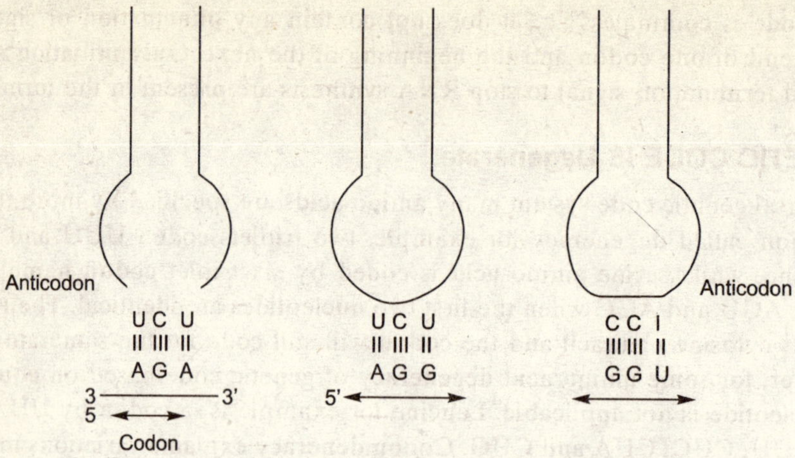

Pairing combination

Base in anticodon (tRNA)	Base in codon (mRNA)
G	U or C
C	G
A	U
U	A or G
I	A, U or C

Table 6.3 Degeneracy of codon

Amino acid	No. of codons	Amino acids	No. of codons
Glu	2	pro	4
Gly	4	Ser	6
His	2	Thr	4
Ala	4	Trp	1
Arg	6	Tyr	2
Asn	2	Val	4
Tsp	2	Leu	6
Cyst	2	Lys	2
Gly	2	Met	1
		Phe	2

Reading of Genetic code : Some of the genes known as suppressor genes causes misreading of a specific non-sense codon. Suppressed genes can reverse the effects of only a small fraction of point mutation that might arise within a gene. They act by reading a stop signal. There are three well characterized suppressor genes that suppress UAG codon. One of the suppressor gene inserts serine, another glutamine and third tyrosine at nonsense position. The suppression of UAA nonsense codon is also mediated by mutant tRNA.

6.5.4 INITIATION CODON

In the concept of central dogma of molecular biology, a specific signals are necessary for the initiation of polypeptide chain at the proper point. The initiation signal for the synthesis of protein is the AUG codon. This AUG codon perform two function in bacteria. When AUG is present at the beginning of messenger RNA (starting codon) it will code for the incorporation of N-formyl methionine (f-met) and AUG is present in any other position of the same messenger RNA will code for normal methionine (f-met). In bacterial cell, a unique fmet tRNA that is different from the tRNA for normal methionine. In eukaryotes initiation codon is also AUG where it codes for methionine. In general AUG is the initiation codon in both prokaryotes and eukaryotes. GUG is an alternate initiation codon for methionine is prokaryotes. However, there are two transfer RNA for methionine, but only one of its commonly used one is the initiation of tRNA. Invitro studies reveals that f-met tRNA can bind and initiate synthesis at GUG as well as AUG codon. Codes for valine amino acid and would require an sort of unusual Wobble in the codon-anticodon interaction. In addition, UUG and CUG codons also initiate initiation. But their involvement in initiation is more rarely than GUG.

6.5.5 TERMINATION CODON

Once translation begins how long would it continue ! There is always a definite check point corresponding to sequence on every gene. At this point fate of polypeptide chain come to an end. There are three termination codons present on mRNA. They are UAG, UAA and UGA favours termination of polypeptide synthesis. During elongation process when ribosome reaches termination codon on mRNA. The completed polypeptide chain is released due to hydrolytic cleavage unlike all other codon, UAG, UAA and UGA codons are not recognized by tRNA. However, these are recognized by special proteins known as releasing factors of protein synthesis. The termination codon were at first called non-sense codon but they were ultimately called as termination codon or signals.

6.5.6. THE GENETIC CODE IN MITOCHONDRIA

Although genetic code is universal, some exceptions can be seen in the genetic code of mitochondrial DNA. They can influence the codon involved in initiation or termination. Exception to the genetic code occurs in the mitochondria of several species. Based on evolutionary studies it was established that there was a universal code that was changed at various points in mitochondrial evolution. One of the most striking features of mitochondrial genetic code is the use of UGA to code for tryptophan. This UGA is actively a termination codon of nuclear genetic material. For example in prokaryotic mycoplasma, UGA is not used for termination but instead codes for tryptophan. In some unicellular protozoans read UAA and UAG, which codes for glutamine instead of termintion codon. In another ciliate, UGA codes for cysteine. Only UAA is used as a termination codon and UAG is not found (Table 6.4)

Another striking feature of mitochondrial code is the presence of only 22 different tRNA. According to Wobble hypothesis a minimum of 31 tRNAs are required to recognize for each codon family and one tRNA is needed to per codon. But in mammalian mitochondria in which there are only 22 different tRNA. At this situation it can be interpreted that one tRNA recogises all four members of a codon family. Hence, their modification reduces to 23. It is the minimum number of tRNA required to recognize all codons. These many changes found in mitochondria of different species suggest that they have evolved separately.

Table 6.4 Codon usage between nuclear genes and mitochondrial genes

Codon	Nuclear genes	Yeast mitochondrial genes	Mammalian mitochondrial genes
CUU	Leu	Thr	Leu
CUC	Leu	Thr	Leu
CUA	Leu	Thr	leu
CUG	Leu	Thr	leu
UGA	Ter	Trp	Trp
AUU	Ileu	Ile	Ile Initiation
AUA	Ileu	Met	Met
AUG	Met	fmet	Met
AGA	Arg	Arg	Ter
			Ter

SUMMARY

Gene is a functional unit of heredity or in general unit of heredity. All the genes need not code of protein but code of rRNA and tRNA synthesis. Hence, genes are the DNA sequence produce RNA. There are organized in chromosomes. Genes an made up of DNA but never from protein. Gene is also a unit of heredity and recombination.

Gene is a shortest segment of chromosome. The subunit of gene is referred as recon. Sequence in gene can undergo alteration referred as mutan gene is a unit of function, also known as cistron. Gene can undergo duplication during its replication. Gene duplication takes place during interphase after mitosis. Genes can be mapped by restriction cleavage. Genes are categorized as dominant gene and recessive gene at the back drop of classical genetics. A dominant gene is the one which suppress the expression of the recessive gene.

Eukaryotic genes are well characterized by the presence of non-coding sequence or introns. Many genes contain more introns than coding exons. Introns can be spliced by spliceosome apparatus exist inside the eukaryotic cell. Interrupted or introns are otherwise called as split genes.

Prokaryotic genes are organized as operon where genes are clustered or adjacent in the same region on chromosome. Number of genes differ from one organism to another. There is linear increase in terms of gene number in the course of evolution of life. A simple influenzae virus contain 12 genes. Bacteria like E. coli contains 2000-3000 genes. Whereas human DNA contains approximately 100,000 genes.

Bacterial genes are in action when its promoters are occupied by RNA polymers. Bacterial promoters are signified by the core conserved sequence like TATA box whereas eukaryotic system it is dominated by CAAT box ac box in addition to TATA box. Eukaryotic genes are organized in chromosome which are then associated with histone protein and forms nucleosomes. Eukaryotic DNA size ranges between < 108 bp to 1011 bp. Chromosomal DNA are characterized by repetitive sequence or tandom repeats. Genetic code is universal. The information molecule DNA passes information to RNA and from their translated to produce proteins. This nature of flow of genetic information could justify the concept of central dogma of molecular biology. Deciphering of genetic code is one of the remarkable achievement in the field of molecular biology. The genetic information is stored in the form of nucleotide triplets also known as triplet codon. Organisation of triplet codon through light on how four-letter language of nuclcic acid is translated into twenty letter language protein.

Codon is the hereditary units that contain information coding for one amino acid. Genetic code explains several salient features. The code is commaless. It does not contain any punctuation or signal required to indicate the end of one codon and beginning of next the genetic code is degenerate i.e., there is more than one code word for most of the amino acids ex. There is single code for tryptophan where as two fold degeneracy for amino acids like glutamine, tyrosine, histidine etc.

Third base is the codon is less specific than the first two. The degeneracy of codon involves only the third base in the codon. Initiation signal for the synthesis protein is the AUG. It is present is at beginning of messenger RNA. It is the starting codon. There are three termination codon like UAG, UAA and UGA favours termination of protein synthesis.

The genetic code in mitochondria slightly varies from universal genetic code. In mitochondrial genetic code use UGA to code for tryptophan. This is termination codon in nuclear DNA.

QUESTIONS

1. Special gene organisation-how do they correlate in plants and animals. (BU May 1996)
2. Give an account of Eukaryotic genes. (BU Nov. 1999)
3. Compare the distinguishing features of prokaryotic and eukaryotic gene.
4. Explain the salient features of gene.
5. What is genetic code ? Discuss its features.

SHORT QUESTIONS

1. Prokaryotic gene (BU Nov. 1999)
2. Eukaryotic gene
3. Plant gene structure
4. Split genes
5. Repetative DNA
6. Operon organization.
7. Codon usage
8. Termination codon
9. Mitochondrial genetic code
10. Wobble hypothesis.

7
CHAPTER

Transcription

Transcription is a process in which RNA is synthesised using one of the DNA strand as template. RNA synthesis is catalysed by the active participation of an enzyme RNA polymerase. Transcription is the first step in gene expression, which start when RNA polymerase tightly associated with DNA at specific sequence of the gene. These are called promoters. RNA polymerase is then move along with template DNA and synthesise RNA until it reaches a termination sequence.

Every living cell has to produce proteins, enzymatic proteins and RNA in order progress smooth passage of all metabolic functions. Lack of expression of any specific set of gene(s) may be detrimental to the cell or organism in general. Thus transcription is a vital process in the growth and development of an organism. Transcription is a highly coordinate process in which the complete association between variety of proteins takes place. Further, DNA-RNA interaction, protein-DNA and protein-protein interaction are indispensable during eukaryotic transcription. Even a single organism like bacteria are able to transcribe all of its genes in a short period of life.

The complete success in transcription is due to the active participation of RNA polymerase and it takes all the credit in this vital process. There are some fundamental differences in the mode of transcription in prokaryotes and eukaryotes.

7.1 MECHANISM OF TRANSCRIPTION IN PROKARYOTE

The process of transcription begins with recognition of particular sequence on a DNA called promoter region by enzyme RNA polymerase. The enzyme is then bind to double stranded DNA with the help of its subunit called σ (sigma) factor. RNA polymerase recognize specific region on DNA called promoters. (Details on promoter will be discussed later) and initiate RNA sequence immediately (Fig. 7.1).

The region of promoters on DNA are the landing site for RNA polymerase, bind tightly and causes duplex to melt and unwind to form transcription bubble. RNA syn takes place within a transcription bubble. At this junction duplex DNA is separated into single strand and

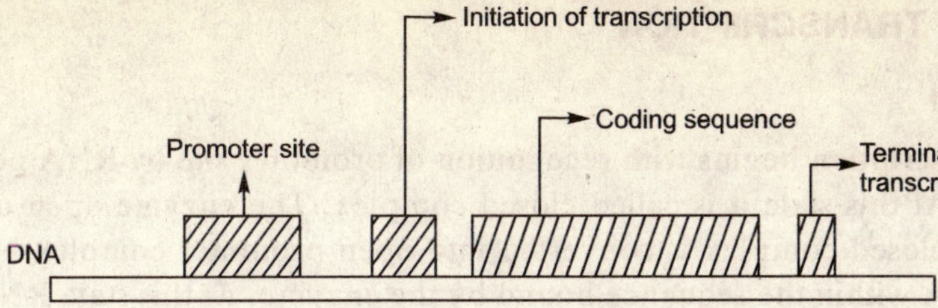

Fig. 7.1. *Overview of Transcription.*

RNA Polymerase

One of the most extensively studied polymerases are E. coli RNA polymerase. There are 7000 RNA molecule are present in E. coli cell. Bacterial RNA polymerase (Mol. Weight 450,000) comprises five subunits. They are two α subunit (Mol. Weight 40,000), one β subunit (Mol. Weight 150,000) one β′ subunit (Mol. Weight 155,000) and one σ subunit (Mol. Weight 70,000).

The β and β′ subunit together constitute catalytic center. The holoenzyme ($\alpha_2\beta\beta'\sigma$) can be separated into two components. One is the core enzyme which contains two α, β and β' another is sigma factor (σ polymerase chain). The α subunit is required for the assembly of core enzyme. The α subunit play a key role in recognition. One of the key function of sigma factor is to recognize and bine RNA polymerase to the promoter site. Hence, a total diameter of RNA polymerase would be around 100Å. Initiation of transcription is performed by σ factor, and rest of the elongation process is done by core enzyme i.e., α_2 β β′. The genes for E. coli RNA polymerase subunits are found in three opersons (Fig. 7.2).

Table 7.1 Bacterial RNA polymerase and other factors

Subunit	Gene	Mol. Weight	Subunit number	Function
α (alpha)	rpo A	40,000	2	
β' (beta′)	rpo C	150,000	1	DNA binding
β (beta)	rpo B	155,000	1	RNA polymersiation
σ (sigma)	rpo D	65,000	1	promoter recognition
ω (omega)	–	12,000	1	
ρ (rho)	rho	–	6	Termination
nusA	nus A	–	1	Elongation

Presence of core sequence in a promoter are the main attraction site for RNA polymerase and are able to associate tightly to initiate transcription.

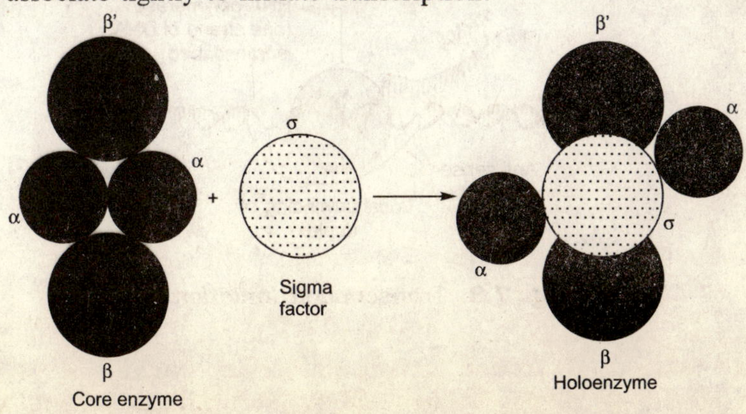

Fig. 7.2. *Prokaryotic RNA polymerase.*

7.1.1 STEPS IN TRANSCRIPTION

7.1.1.1 *Initiation*

Initiation of transcription begins with recognition of promoter site by RNA polymerase and bind loosely to it. At this state it is called closed complex. The enzyme open up duplex into single strand. The closed complex is converted into open promoter complex by melting of a short region of DNA within the sequence bound by the enzyme. At this state RNA polymerase increases its affinity and bind much more tightly. Presence of sigma factor in RNA pol ensures stable binding of enzyme to DNA at promoter sitee. Only holoenzyme of RNA polymerase (includes sigma factor) can initiate transcription. Whereas the core enzyme (without a sigma factor) cannot perform initiation as it fails to recognize sequence between promoter and other sequence of DNA. Thus, it is only a sigma factor comes to the rescue during initial transcription process. The sigma factor in holoenzyme ensures tight binding at promoter for several hours. The promoter region consist of unique six nucleotide sequence. These are known as conserved sequence. These occur at about 10 and 35 nucleotide pairs before transcription initiation site.

RNA polymerase unwinds DNA continuously as it extends the growing chain. Thus exposing the template strand against which RNA precursors are polymerized. At about 10-12 nucleotides of the growing RNA chains are hydrogen bonded to DNA at any one time. A short chain of 2-9 bases of RNA is synthesised and released (Fig. 7.3) then continue further synthesis of RNA. Beginning of RNA synthesis is immediately followed by the release of sigma factor from the holoenzyme but the core enzyme on DNA continue to synthesis RNA. Thus sigma factor is required only to recognize promoter, and the rest of holoenzyme shoulder responsibility in RNA synthesis. The σ factor once released from transcription site are available to transcribe other genes by binding to holoenzyme.

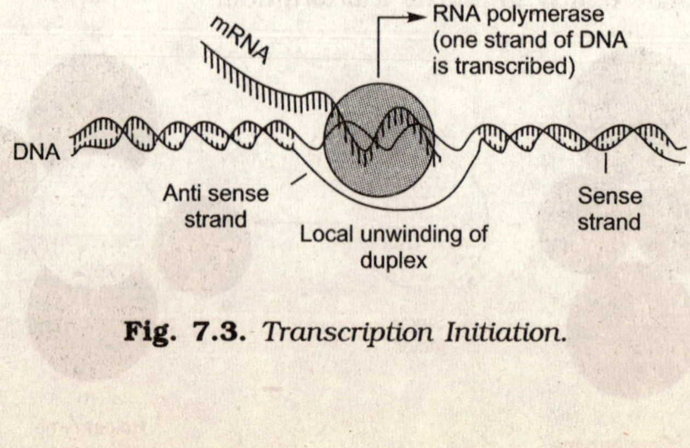

Fig. 7.3. *Transcription Initiation.*

7.1.1.2 *Elongation*

Elongation step is the lengthening of RNA chain. Once the first ribonucleotides are polymerized, core enzyme will move along with DNA, continuously separating two strand. Separation of DNA strand ensures continuous incorporation of ribonucleotide. Finally RNA chain elongates. In the later process newly synthesised RNA is displaced from the template DNA strand. The renaturing of DNA into double helix takes place continuously follows RNA synthesis. The use of radiolabelled ATP, GTP, CTP and UTP in the experiments showed that RNA chains have a purine at their 5′ end i.e., they begin with a pppA or ppG thus, RNA chain usually initiate with pppA or ppG. Further confirmation reveals that RNA synthesis takes place in 5′–3′ direction. The genes are transcribed continuously until RNA pol. Reach the destiny of termination.

7.1.1.3 *Termination*

Termination is the last stage in the process of transcription where in RNA synthesis will terminate and release from the template DNA. There are specific sequences on template DNA known as termination sequence. Once RNA polymerase facilitate transcription. The enzyme move along sense (template) strand of DNA simultaneously synthesing RNA chain until it approaches its last destiny i.e., termination site. When RNA pol approached termination site it stops adding nbhl to the growing RNA chain. It is then released from DNA template. At least two termination sites have been identified in E. coli. One is a specific sequence known as intrinsic terminator in which the core enzyme a RNA polymerase can terminate transcription in absence of rho factor. The second one is the rho factor. It is a protein play a indispensable role in termination. Several mutation studies confirmed its role in transcription.

7.1.1.4 *Intrinsic Termination*

There are two special feature are associated with intrinsic terminator. One is the hair pin feature in the secondary structure of RNA and other is run of U residue. Presence of these two features are very much essential for termination of transcription. Hairpin feature usually contains G-C rich region near the base of the stem.

7.1.1.5 *Rho (p) Factor*

Rho (p) factor is basically a protein attains molecular weight of 5,000 Daltons, committed to perform termination. It acts as a helper protein for RNA polymerase in terminating RNA synthesis on the template RNA. There are 60-90 base pair long Rho factor dependent terminator sequence, usually rich in 'C' residue. According to proposed model Rho factor is first attached to RNA molecule and moves quickly along the chain and reached the termination sequence. The enzyme polymerase pause when it reaches termination site and termination takes place if already rho factor is present on the termination sequence. In the absence of RNA polymerase rho factor can exhibit 5′–3′ helicase action that can break hydrogen bond between DNA-RNA hybrid at the expense of ATP energy and release the RNA chain. It is followed by renaturation to form Duplex DNA (Fig. 9.4 and 9.5).

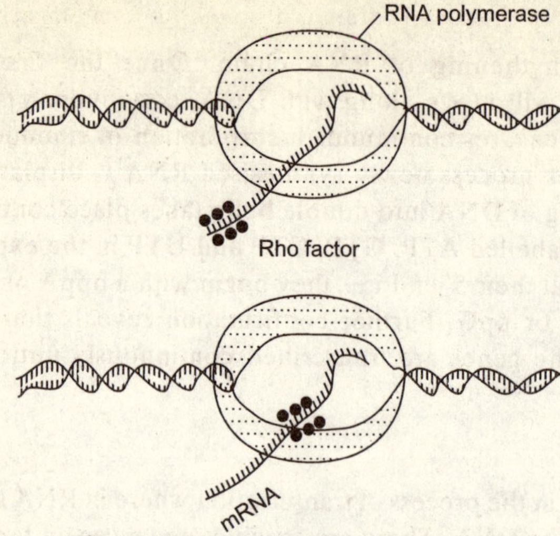

Fig. 7.4. *Termination of transcription.*

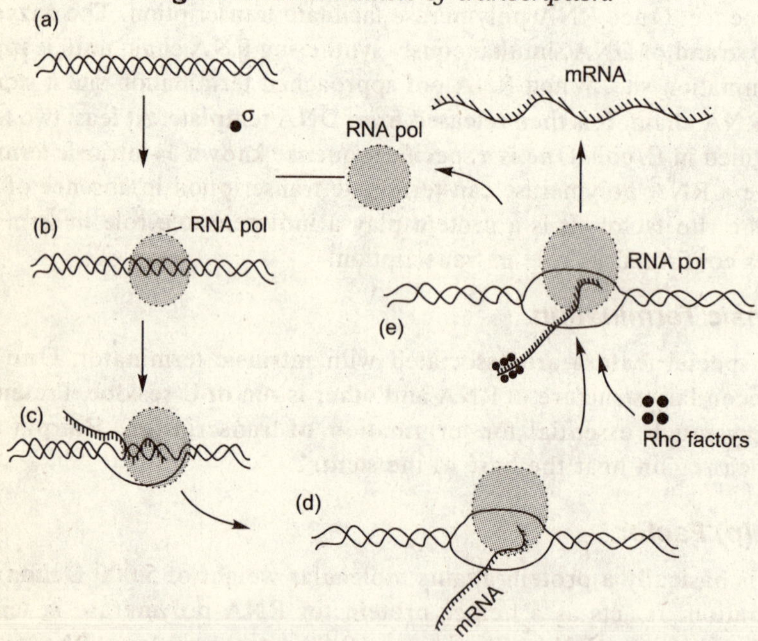

Fig. 7.5. *Overview of transcription.*

7.2 TYPES OF TRANSCRIPTS

Transcription of one of the strand of template DNA resulted in the synthesis of RNA molecules like mRNA, tRNA and rRNA. In prokaryotes, transcription genes for all types of RNA (mRNA, tRNA, rRNA) are performed by only one RNA polymerase. The messenger RNA carries information that specifies insertion of particular aminoacids in the polypeptide chain. The initial primary transcript are unsuitable to undergo translation. Therefore, certain modifications are required essentially before undergo translation. Certain interruptive sequences are present in the DNA which does not codes for protein are called as introns or non coding sequence. These are predominantly present in eukaryotic DNA. Since, transcribed RNA contains introns, these are required to eliminated by mRNA processing.

All the cells are well equipped with post transcriptional processing machinery. In prokaryotes a long mRNA molecules are produced by transcription of several genes organized as operon. The coding sequences are separated by short stretch of spacer region in DNA, which can transcribe but not translated which are removed during mRNA processing. In addition, prokaryotes contain termination sequence predominantly represented by U sequence known as attenuator, may code for short polypeptide. Final processing of mRNA may assume secondary structure like hair pin loop. In prokaryotes, the rRNA and tRNA are transcribed as precursor which are longer than the functional RNA which will have to undergo endo and exonucleolytic cleavage in the final processing to assume their final size. The rRNA and tRNA undergo enzymatic modifications like methylation with the help of methyl transferase. In bacteria, the mRNA is transcribed and translated in a single cell compartment. The stability of mRNA in a cell is a matter of concern. The prokaryotic mRNA is less stable than its eukaryotic partner. It is degraded in 5′-3′ direction after translated by ribosome in a rapid succession.

7.3 EUKARYOTIC TRANSCRIPTION

In eukaryotic system transcription takes place within the nuclear compartment followed by translation of mRNA in cytoplasm. Translation of mRNA can begin before its synthesis is completed. The mRNA will have to travel from nucleus into the cytoplasm for translation. Actually translation begins simultaneously when one end of mRNA molecule pass through the nucleus into cytoplasm and ribosome get attached to it.

In the initial phase of transcription the heterogenous nuclear RNAs (hnRNA) partly hydrolysed by nucleases within nucleus. These are partly converted into mRNA. Primary transcript of mRNA contains interrupted noncoding region known as introns. These are removed during maturation of mRNA. Post transcriptional modifications are specialized process in eukaryotes, involves additional modifications like capping and poly A tail attachment. Unlike bacterial polymerases, eukaryotic polymerases transcribe the genes slightly in a different manner. They require special accessory factors to initiate the process.

7.3.1 Eukaryotic RNA Polymerases

Complexity of eukaryotic transcription could be attributed to RNA polymerases and transcription factors. In eukaryotes there are three different types of RNA polymerases can be seen. They are RNA pol I, pol II and pol III. These DNA dependent RNA polymerase shoulder the responsibility of transcribing protein coding genes, tRNA genes and rRNA coding genes, RNA pol I transcribe rRNA . RNA pol II transcribe mRNA genes and RNA pol III are able to transcribe tRNA and other small RNA coding genes. These three polymerases are localized in nuclei of eukaryotic cell.

7.3.1.1 *RNA pol I*

RNA polymerase 1 transcribe genes only for ribosomal RNA. This enzyme is localized in the nucleolus can produce almost about 70 per cent of the total cellular RNA. It transcribe a long transcript containing 18S, 5.8S and 28S genes of DNA. RNA pol I requires two supportive factors known as UBF_1 polypeptide binds to SL1GC rich in the core promoter. Once both factors are bound (transcription factor) RNA polymerase I combines with core promoter to initiate transcription.

7.3.1.2 *RNA pol II*

RNA pol II can be localized in nucleoplasm, responsible for the synthesis of heterogenous nuclear RNA (hnRNA), which is a precursor of mRNA. This enzyme is inhibited by low concentration of a-amanitin. RNA pol II has more than 10 subunits. The largest subunit in polymerase has a carboxyl terminal domain (CTD) these are characterized by repeated sequence of 7 aminoacids. However, number of aminoaicds repeat are different among organisms.

7.3.1.3 *RNA pol III*

Nucleoplasm is the site of RNA pol III activity. It exhibit minor enzyme activity. This enzyme synthesis about 10 per cent of cellular RNA especially tRNA and other small RNA.

All eukaryotic RNA polymerases are aggregated proteins of 530,000 Daltons. They consist of 8-14 subunits. The major component are two large size subunits of molecular weight 220,000 and 140,000 and several smaller subunits. These smaller ones are believed to be involved in the formation of phospho-diester bonds.

7.3.2 Mechanism of Transcription

In general, mechanism of transcription is similar to prokaryotes. In comparison to prokaryotic promoter, eukaryotes comprisers of several promoters which are organized into 30 nucleotide upstream from the site of transcription. The A=T rich sequence known as TATA box is a promoter situated for away from the site of transcription i.e., - 25 base pair region. These are recognized by eukaryotic pol II. Some of the proteins are present even far away from the site of transcription. Between (– 40 and 110) called CAAT bix and and GC box which are recognized by RNA pol II. Transcription initiation in eukaryotes requires many accessory factors like transcription factor protein. These proteins are able to interact with DNA and bind to promoter region or enhancer region. (Specialised sequence on RNA situated far away from transcription site). Some of the factors may recognize RNA polymerase or may be incorporated into an initiation complex only in the presence of several other proteins. A significant difference between transcription of eukaryotic and prokaryotic mRNA is that initiation at eukaryotic promoter involves a large number of transcription factors that bind to a variety of cis-acting element. Transcription factors bind to upstream sequences influence transcription initiation. Other factors in the basal apparatus are able to contact directly or indirectly. However, it does not involve contact, with RNA polymerase, making protein, protein contact with general transcription factors and promoters. It participate in the formation of basal apparatus.

The common features of regulation of eukaryotic transcription is positive. Eukaryotic transcription unit generally contain single gene and termination occur beyond the end of coding region. The termination of RNA chain synthesis takes place by polymerase I and III at discrete sequences in defined reach. However, the nature of termination by pol II is not celar.

Initiation of transcription begins with the complex formation at promoter region. The first step in complex formation at TATA region of promoter is the binding of factor TF II D. It is followed by joining of TF II A to the complex. The transcription factor TF II F itself do not bind directly to the complex. TF II F brings RNA polymerase to the assembly site and finally joins the complex. In addition two more factors TF II E and TF II H joins the transcription complex (Fig. 7.6). The helicase and ATPase activity of TF II H helps in the movement of RNA polymerase II from transcription initiation site. Transcription begins with polymerase movement followed by release of all transcription factors from the DNA.

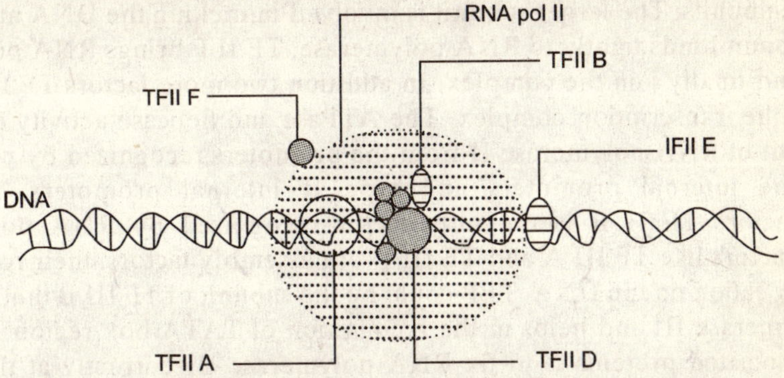

Fig. 7.6. *General view of Transcription in Eukaryote.*

Eukaryotic transcription is a highly coordinated process which involves participation of the enzyme RNA polymerase and proteins. RNA polymerase enjoys all privileges in prokaryotic transcription as they do not require the support of any proteins. However, the situation is different in a eukaryotic system. RNA polymerase alone cannot perform the transcription. Hence certain group of proteins known as transcription factors are indispensable to initiate the process. RNA polymerase plus transcription factors together called basal transcription apparatus forms a stable complex at transcription initiation site by binding to TATA region in the promoter sequence of DNA.

Transcription factors are generally divided into general transcription factors (GTF) and sequence specific transcription factors (SSTF). General transcription factors are indispensable to transcribe all RNA pol II dependent genes. Where sequence specific transcription factors regulate transcription. There are six general transcription factors (GTF) are required in order to transcribe RNA polymerase dependent genes. These are TF II A, TF II B, TF II D, TF II E, F and H.

Transcription Factors for RNA Polymerase II

Transcription factors for RNA polymerase II (TF II A, TF II B, TF II D, TF II E, F and H) are basically DNA binding protein and protein interactive polypeptides. TF II D transcription factor can recognize the promotor that contains TATA sequence and binds specifically at this region. TF II D contains two types of components TATA binding protein (TBP) and another subunit TBP associated protein or factors (TAF). TBP is a small protein of 30 kD which recognizes TATA box in the promoter. TF II D contains various TAFs that can recognize different promoters. The total mass of TF II D is 800 kD. The mass of the components of TF II D i.e., TBP and TAF varies from 40-240 kD. TBP is associated with TAF and binds to the RNA at promoter region. TBP has a unique property of binding to the minor grove of DNA. The inner surface of TBP binds to DNA whereas its large outer surface makes contact with other proteins involved in transcription.

Initiation of transcription begins with the complex formation at the TATA promoter region. The binding of TF II D factor (Association of TBP with TAF binds to TATA box). This is followed by the joining of TF II A to the complex. After binding TF II A is activated by TBP. The complex is able to protect a region. Binding of TF II B takes place on both upstream and downstream region (–10, + 10) on DNA. It binds adjacent to TBP and looses its association with DNA. The transcription factor TF II F itself will not bind directly to the complex. It consists of two subunits. The larger subunit is involved in melting the DNA at initiation point. The smaller subunit binds tightly to RNA polymerase. TF II F brings RNA polymerase to the assembly site and finally join the complex. In addition two more factors TF II E and TF II H get attached to the transcription complex. The ATPase and helicase activity of TF II H helps in the movement of RNA polymerase II from the promoters recognized by pol III are of two kinds, type one internal promoters and type II internal promoters. Among several transcription factors TF II B is the only factor recognized by RNA pol III to initiate transcription. Factors like TF III A and TF III C are assembly factors, their role is to assist TF III B to a right location on the DNA. TBP could be the subunit of TF III B that interact directly with RNA polymerase III and helps in the recognition of TATA box region. The function of TBP and its associated proteins is to fix RNA polymerase III correctly at the start point of transcription.

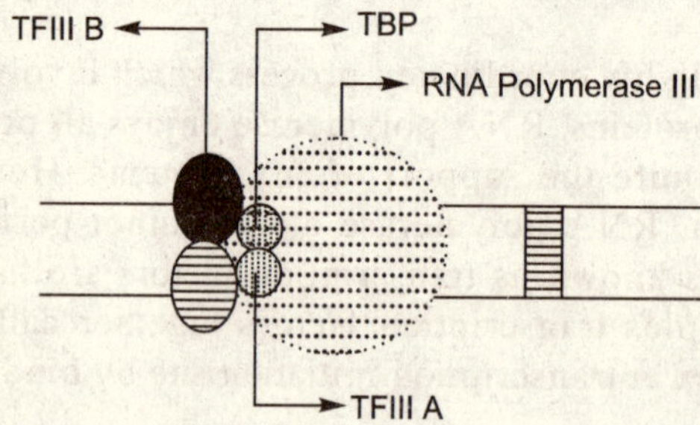

Fig. 7.12. *Assembly of transcription factor and pol III.*

Transcription Factor for RNA Polymerase I

RNA polymerase I transcribe genes for ribosomal RNA. It requires two helping factors, one is UB F 1, a single polypeptide that binds to GC rich region of the core promoter and another is SL 1 which can bind to the DNA. The RNA polymerase I can bind to the core promoter only after the binding of these two helping factors. Once the assembly takes place (UB, F, SL 1 and pol I) transcription takes place.

Discrete group of mammalian transcription factors contain domains which are typical of those found in many DNA binding proteins. At least two of the domains known as Leucine Zipper (LZ) and helix-loop-helix (HLH) have been recognized and extensively studied. These two domains HLH and LZ are predominantly present next to the region of the positively charged amino acid sequence. Invitro manipulation studies have shown that DNA binding regions are specifically located in these positively charged amino acid region.

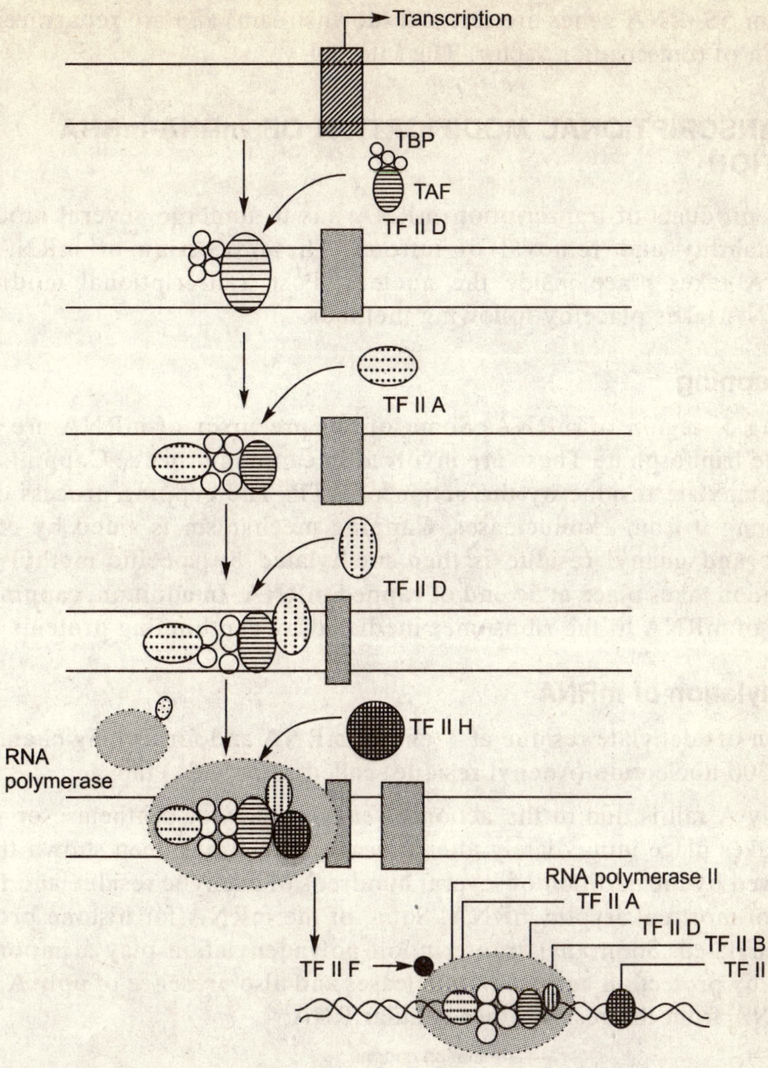

ENHANCER ELEMENT

Enhancer element is a DNA sequence that increases the amount of RNA synthesis. It has a strong influence on transcription by RNA polymerase II. It can effectively help transcription independently of its position and orientation with respect to the initiation. In otherwords activity of promoter at least in some cells is substantially increased by the presence of enhancers, whether it is situated in the upstream or downstream direction or far away from the initiation site. Transcription begins with polymerase movement, followed by release of all transcription factors from the DNA.

RNA polymerase III transcribe tRNA genes like 5S genes. It requires few transcription factors to initiate transcription. Some of the transcription factors loike TF II A and TF II B can associate with pol II and polymerase III for the transcription of both protein coding and tRNA genes.

The promoter for 5S tRNA genes are internal (downstream) and are recognized by pol III with the association of transcription factor. The internal

7.4 POST-TRANSCRIPTIONAL MODIFICATION OF mRNA-mRNA MATURATION

In eukaryotes the products of transcription mRNA has to undergo several modification in order to attain stability and removal of introns. The processing of mRNA precursor heterogenous RNA takes place inside the nucleus. Post transcriptional modifications or maturation of mRNA takes place by following methods.

7.4.1 mRNA Capping

It involves capping 5′ region of mRNA. Some of the precursor of mRNA are adenylic or guanylic nucleotide triphosphate These are involved in capping process. Cappint at 5′ end by condensation of guanylate residue by the action of GTP. The capping process can stabilize mRNA by protecting it from exonucleases. Capping mechanism is aided by enzymes like guanyl transferase and guanyl residue is then methylated by specific methyl transferase. Extensive methylation takes place at 5¢ end of capped mRNA. In addition, capping of mRNA promotes binding of mRNA to the ribosomes mediated by cap binding proteins (Fig. 7.7).

7.4.2 Polyadenylation of mRNA

It involves addition of adenylate residue at 3′ end of mRNA and forms long chain of poly (A) consisting of 200-300 nucleotide (Adenyl residue) called as poly (A) tail.

Addition of poly A tail is due to the action of enzyme poly A synthetase or polymerase. Polyadenylation takes place immediately after transcription. It has been shown that cleavage of mRNA is followed by the addition of several hundreds of adenylic residue and forms a poly (A) tail at 3′ end of most eukaryotic mRNA. Some of the mRNA for histone protein do not generally polyadenylated. Soon after transcription polyadenylation play a important role in stabilizing mRNA by protection against exonucleases and also presence of poly A helps in the separation of mRNA from the rest of other cellular RNA.

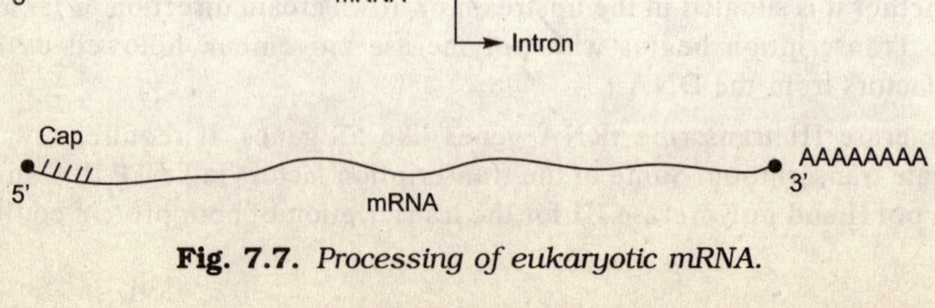

Fig. 7.7. *Processing of eukaryotic mRNA.*

7.4.3 Removal of introns (RNA splicing)

In prokaryotes the sequence of amino acid in a protein chain is exactly that of codons in the messenger RNA. However, in eukaryotes protein is not exactly to that of codons in the messenger RNA. The DNA of eukaryotic system contain several interruptive sequence do not code for proteins. This non-coding sequence are called as introns. They may be present within the coding sequence. The primary transcription product of split genes contain both introns and exon sequence (coding sequence) and constitute pre-messenger RNA (pre-mRNA). The introns are long more than 1 kb and exons are relatively short i.e., 100 to 200 base pairs. Following capping are polyadenlyation, pre mRNA splicing takes ploace in the mRNA to remove introns before they enter into the cytoplasm. The precursor mRNA undergo third phase of maturation in which introns are completely eliminated. The process by which introns are removed are called as RNA splicing. Several splicing mechanism have been observed in the elimination of introns. They are removed of introns by lariat formation and by spliceosomes.

7.4.3.1 *Removal of Introns through Lariat*

In this mechanism of RNA splicing, first a cut is made at the 5′ splice site (Fig. 7.8) separating left exon and right intron. The left exon appears as a linear molecule. The right intron exon molecle assumes lariat or loop structure in which 5′ terminus will be generated at the end of introns become linked by 5′-3′ bond to a base within the intron. The target base 'A' in a sequence called the branch site.

In the second stage cutting at the 3¢ splice site releases the free introns in lariat or loop form. While the right exon is joined to left exon. Loop cariet is then debranched and converted into linear form. Excised introns are then degraded immediately (Fig. 7.8).

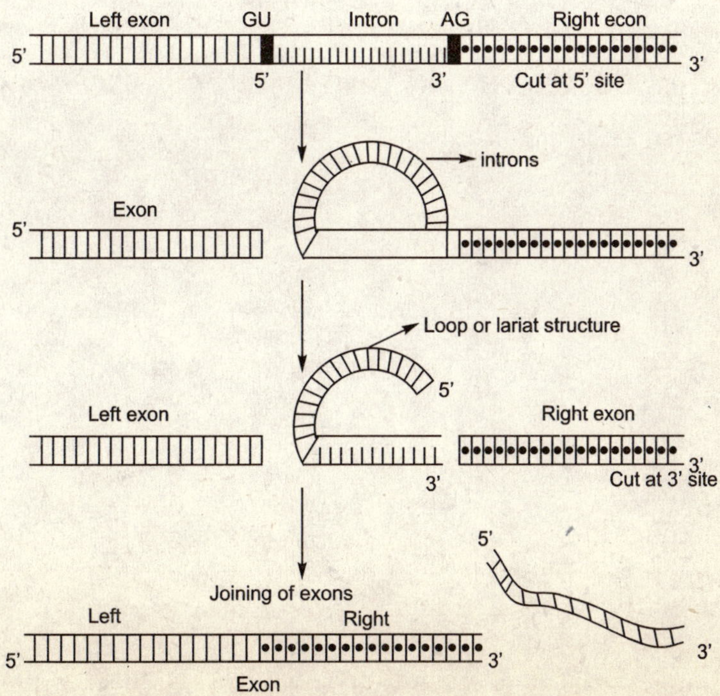

Fig. 7.8. *RNA splicing by Lariat.*

7.4.3.2 *Removal of Introns by Spliceosome*

The eukaryotic call can also perform splicing of introns with the aid of spliceosome apparatus. The splicing apparatus contains both proteins and RNA. As eukaryotic cell no longer can tolerate the presence of introns, these two helper molecule swung into action to eliminate non coding sequence in the gene. There are number of small RNA molecule found in eukaryotes. Those found in cytoplasm are called SnRNA molecule and those found in cytoplasm are called ScRNA (small cytoplasm rRNA). They exist as ribonucleoprotein particle (SnRNP). The complex called splicesome contains SnRNP associated with some proteins. It consist of 50-60 ribonucleoprotein particle. Some of the SnRNPS involved in splicings are U_1, U_2, U_4, U_5 and U_6. Splicing is generally done by recogising consensus sequence first and spliceosome complex assembles at this site. It is followed by the clevage of intron site and ligation of exons. Recognition of consensus sequences is done by the both RNA and proteins. (Fig. 7.9)

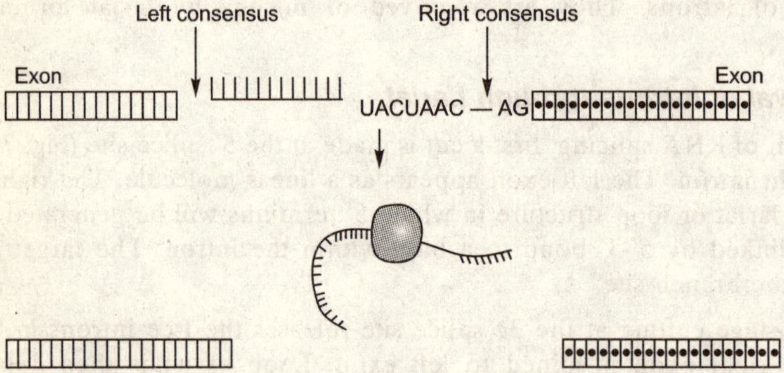

7.5.1 Promoters of Prokaryotes

Among prokaryotes promoters of E. coli have been studied in detail. Bacterial promoter region is about 40 bases long with include common sequences of about 6 bases which is located upstream from the site at which RNA synthesis begins. This region is called "Pribnow" box. The pribnow box centers around the –10 position. The –10 sequences is hexamer, its consensus sequences are TATAA. All the promoters contains sequence similar to TATAA at –10 consensus sequence. Similarly, another consensus sequence is TGTTGACA, centers around –35 i.e., about 35 base pair before the site of transcription. This region is critical for the rapid initiation of transcription for most bacterial genes. Invitro experiments reveals that T_4 phage s factor binds primarily at –10 region. The –10 and –35 hexamer of a promoter are the effective region. Where RNA polymerase binds to these site. Certain sequences in the upstream DNA sequence can increase promoter activity. Natural sequences between 50 to 150 nucleotide upstream of the start site for ribosomal RNA synthesis are essential for complete promoter activity.

–35 –10

TTGACA TATAAT

AAATGAGCTG [TTGACA] ATTAATCATCGAACTAG [TTAACT] AGTA CG CAAGTTCA

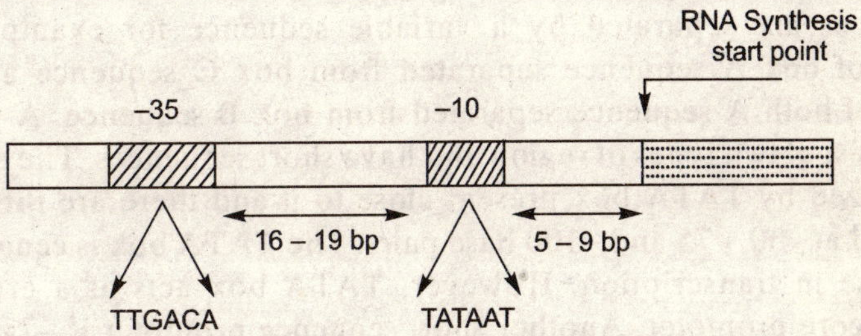

The presence of essential nucleotide in all promoters is said to be conserved sequence. The short strech in a promoter is essential for its function. There are four conserved sequences in bacterial promoter (a) the start point (b) the –10 sequence (c) the –35 sequence (d) sequence in the separation between –10 and –35 sequences.

Presence of –35 sequence in a promoter acts as a perfect signal for recognition by RNA polymerases. The conserved sequence of the –10 site consist exclusively of A–T base pairs. The AT rich region facilitate melting of DNA into single strand as it demands low energy for strand separation.

Gene	Factor	– 35 sequence	– 10 sequence
rpoD	σ^{70} (sigma)	TTGACA	TATAAT
rpoH	σ^{32} (sigma)	CCCTTGAA	CCCGATNT
Flir	σ^{f} CTGGNA	CTGGNA CTA	TTGCA

7.5.2 Promoters of Eukaryotes

Due to complexity of genome there is tremendous pressure on the selective turned on of genes resulted in differential gene expression. Several types of promoters have been identified among eukaryotes. There are few contrasting difference between prokaryotic promoters from eukaryotic promoters. In eukaryotes TATA box is situated in the upstream direction far away from the site of RNA synthesis, example is at –25 in yeast. In addition to TATA box, some other sequences situated even further i.e., between - 40 and 110 are called CAAT box and GC box. The core promoters surrounds the starting point extending from –45 to + 20. These are sufficient for transcription to initiate rich in GC base pairs. Promoters are present both before upstream, and down stream direction. RNA polymerase in eukaryotes uses both downstream and upstream promoters for ex. RNA polymerase III uses both downstream and upstream promoters. The promoters for 5S and tRNA genes lie downstream of the start point. The promoters for small nuclear RNA (SnRNA) genes are present upstream of the start point. The promoters for 5S RNA transcript is present between position + 55 and + 80 within gene (signifies down stream).

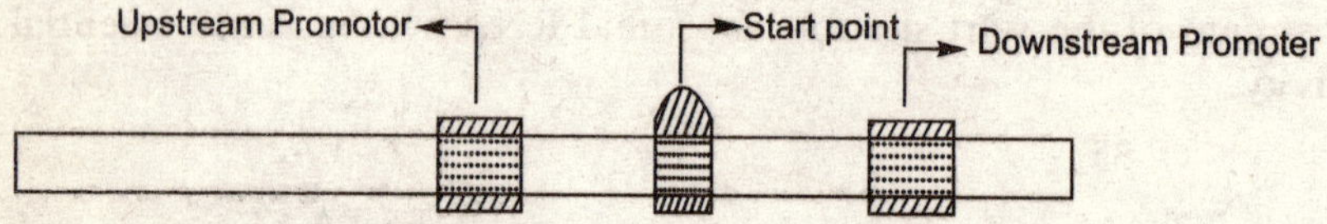

There are three types of promoters for RNA polymerase III of which two are internal promoters. Each containing bipartite structure. Bipartite is a structure of promoter in which two short sequences are separated by a variable sequence for example type 1 internal promoter consist of box A sequence separated from box C sequence and type 2 internal promoter consist of both A sequence separated from box B sequence. A promoter for RNA polymerase II consist of two types of region and have short sequences. The start point synthesis of RNA is recognized by TATA box present close to it and there are three short sequences promoters centered at –30, –75 and –100 base pairs. The TATA box is centred at –30 and play lukeworm response in transcription. However, TATA box acts as a crucial position as a component of the core promoter. Another short sequence promoter is –75 and it is known as CAA box which is named after its consensus sequence. It is often located close to –80. The CAAT box play a significant role in determining the efficiency of promoter. The presence of CAAT box could accelerate promoter strength.

Third important promoter for RNA polymerase is GC box centered at –90 which contains consensus sequences GGGCGG. Multiple copies of these sequences are present in the promoter.

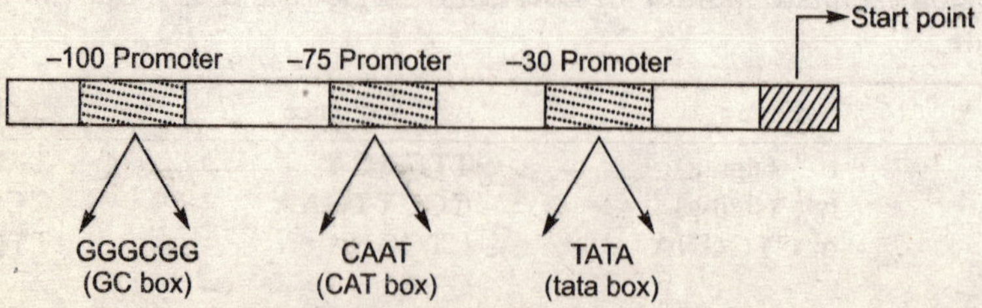

Fig. 7.10. *Eukaryotic promoters.*

The promoter always facilitate transcription provided only in the down stream direction but promoters like CAAT box and GC boxes can able to function in either orientation i.e., transcription can takes place both at upstream and downstream direction. Fourth additional element (promoter !) which aid RNA polymerase bind is known as "octamer" of 8 base pair element. Transcription factors bind to different promoters. The protein C/EBP which binds specifically to CAAT box. TFIID also binds specifically at TATA box on the other side. In the similar fashion transcription factor oct-1 binds to the octomer to activate genes. The GC box is recognized by the factor Sp_1.

7.5.2.1 *Plant Gene Promoter*

More than a decade of research have been carried out for studying plant promoters in gene transcription by RNA polymerase II. The core promoter are minimal promoters that contain TATA box. The TATA box in plants are located about –20 to –25 base pairs from transcription site like those in animals. Several promoters have TATA box that are similar to the consensus TATA sequence found in eukaryotes. Most of the plant promoters residue within 300 bp upstream to transcription site.

7.5.2.2 *Light Responsive Promoters*

Plant perceive certain wavelength of light by certain photoreceptors involved in a kind of signal pathways that regulate photomorphogenetic regulatory cis-acting sequences in light-regulated genes have been identified by analysis of promoters. Elements that are found in light-regulated promoters include GT-1 site, 1-box or GATA box, the G box and AT rich elements in 5′ GGTTAA 3′ are the general feature of light response promoters. This elements was identified in the rbc-DNA promoter. The I box or GATA motif contain sequence GATA (GATA motif) are found in many light-regulated genes promoters. The core sequence 5′-CACGTG-3′ in G boxes is another element that is commonly found in light responsive genes. G boxes are found in many plant genes. Many transcription factors preferentially bind to G-boxes are called G-box binding factors (GBFs).

AT rich motif are found in many plant promoters. AT rich sequences are found in animals and yeast promoters. AT rich sequences in the pea rbc-S–3A and glutamine synthetase gene promoter function as activation elements. On the other hand, a AT rich elements in nicotiana species chlorophyll a/b binding cab-E gene promoter act as a negatively regulated element. It is clear from the study that light responsive promoters generally contain GT-1, GATA and G-boxes that represents key regulatory elements in genes that respond to light.

7.5.3.3 *Heat Responsive Promoters*

Plant and other eukaryotes contain various heat shock responsive genes. The heat shock responsive element in the promoters of all eukaryotes consist of 3 bp (5′-GAA 3′). Heat shock promoter element (HSE) are generally located close to the TATA box aout 15-118 bp upstream region. Some plant heat shock genes contain a CCAAT box located 50-100 bp upstream of transcription site. AT rich sequence are commonly found in heat shock gene.

7.5.3.4 *Seed Specific Promoters*

The promoters of the genes that are activated during seed development have been studied in several plant speices. Many of these genes include seed storage proteins. The seed storage proteins and lectin gene promoter include G-box and E-box, 5′-ACACACGTCAA 3′, vicilin 5′-GCCACCTCAG-3′ and legumin promoter box CATGCATG etc. These cis elements appear to be essential for seed specific expression in seed storage protein genes.

SUMMARY

Transcription process involves synthesis of mRNA, rRNA and tRNA. It begins with promoter recognition by RNA polymerase. There is only one RNA polymerases for bacterial transcription. Among eukaryotes three RNA polymerases are involved in transcription of genes. They are RNA pol I, RNA pol II and RNA III. RNA pol I transcribe ribosomal genes. RNA pol II participate in the transcription of all protein coding genes. Whereas RNA pol III transcribe only tRNA genes on DNA. All RNA polymerase recognize particular gene promoters. In prokaryotes promoters occur at about –10 and –35 nucleotide pairs in the upstream direction of start synthesis. RNA synthesis elongates as the polymerases move along DNA in 5′–3′ direction. Termination of RNA chain takes place at specific sequences. Termination is aided by rho factors.

Post-transcription modification of RNA takes place predominantly in eukaryotic system by adding guanyl residue in a process known as mRNA capping. Polyadenylation takes place by adding poly A tail at 3′ region. Introns are noncoding intervening sequence present in eukaryotic genes they will be removed by certain RNA-protein spliceosome complex. This phenomena is known as RNA splicing.

Promoters of prokaryotes are TATA box present as a core sequence at –10 base pair region and CAAT box at –35 region. Eukaryotes contain several hundreds of promoters. In addition to TATA box some other promoters are situated at –40 and 110 are called CAT box and GC box. There are different promoters for different RNA polymerases. Several plant gene promoters like light responsive (G1-site), heat responsive promoter (GAA) and seed specific G box and E box have been identified. Several transcription factor proteins can bind to these promoter region especially in eukaryotes and initiate transcription.

QUESTIONS

1. Explain the process of transcription in prokaryotes. (BU Nov. 1999) (BU Nov. 2000)
2. What are the post-transcriptional modifications in eukaryotes? (BU Nov. 1999)
3. Explain the mechanism of transcription in prokaryotes. (BU April 2000)
4. Explain the mechanism of mRNA splicing (BU April 2001)
5. What is a promoter ? Explain the features of prokaryotes and eukaryotic promoters.

SHORT QUESTIONS

1. RNA Splicing
2. Eukaryotic RNA polymerase
3. Prokaryotic promoters
4. Eukaryotic promoters
5. TATA box
6. mRNA capping
7. RNA polyadenylation.

8

CHAPTER

Translation–Protein Synthesis

8.1 TYPES OF RNA

Transcription of protein coding and RNA coding genes (rRNA and tRNA) are facilitated by RNA polymerases. Transcription process generates different types of RNA like messenger RNA (mRNA) transport RNA (tRNA) and ribosomal RNA (rRNA).

8.1.1 Messenger RNA (mRNA)

Synthesis of mRNA is due to the transcription of protein coding genes. Transcription and translating are the two indispensable process closely linked and occur simultaneously. The ribosomes can attach to mRNA even before the completion of transcription.

Prokaryotic mRNA are more stable than eukaryotic mRNA. Due to the lack of compartmentalization translation of mRNA takes place within the viscinity of bacterial chromosomes. In contrast compartmentalization of nucleus and cytoplasmic space in the eukaryotic cell would force the mRNA to undergo translation only in cytoplasmic region. In E. coli cells there are 600 different types of mRNA, all together constitute 1,500 moles per cell, at least 2-3 copies were found for each type. mRNA represents only a small percentage (2-4%) of the total cellular RNA. In bacteria the half life of mRNA molecule is about 1-2 minutes. Since mRNA is known for its instability can undergo degradation in 5′-3′ direction i.e., in the same direction as the synthesis of mRNA and their translation. In bacteria, rapid synthesis and degradation helps in adaptation. Eukaryotic mRNA exhibit considerable stability and they can exist for longer time. The half-life of mRNA of mammalian cell in culture ranges from few hours to 24 hrs. Availability of mRNA for protein synthesis may be from few hours to days. The prokaryotic mRNA is synthesised as long polycistronic mRNA i.e., many structural genes are transcribed in mRNA as a single unit. In eukaryotes its transcription product is monocistronic mRNA i.e., single gene into single mRNA. A polycistronic of RNA as a result of uninterrupted transcription of about ten adjacent genes of a operon. Thus it can be 3 times longer than the ribosomes and 120 times lower than the tRNA. All mRNA contain two specific region begins with initiation codon AUG and the termination codon UAG, UAA and UGA. It contain leader region in upstream of initiation codon and downstream direction i.e., 3′ and other is tail region.

8.1.2. Transfer RNA (tRNA)

Transfer RNA (tRNA) is an adapter molecule occupies nearly 10-15 per cent of the total RNA in the cells of E. coli. This RNA is a polynucleotide chairs between 75-95 nucleotides in length and play a important role in protein synthesis as an adapter molecule. In E. coli there are more than 50 per cent tRNA able to utilise 61 codons among possible 64 triplet codons of genetic code. Some tRNA can able to recognise more than one codon. The tRNA can bind covalently to their corresponding amino acid through a sequence, CCA at 3′ end. This entire complex is known as aminoacyl-tRNA. It facilitate particular amino acid to brought to the ribosomal site. The closed clover leaf model of tRNA exhibit following features.

(a) Transfer RNA contains four arms, they are acceptor arm, TψC arm, anticodon arm and the 'D' arm.

(b) The acceptor arm contains CCA at 3′ end which are unpaired. This can be attached to amino acid. This attachment site is known as acceptor end.

(c) TψC arm contains triplet sequence in which modified base known as pseudouridine is present (ψ symbol stands for pseudouridine).

(d) The anticodon arm or loop is present exact opposite from the acceptor end or arm. The anticodon form triplet bonds with mRNA codon.

(e) The 'D' arm, which contains dihydrouridine.

(f) All tRNA exhibit secondary and tertiary structure and share certain common feature.

(g) tRNA assumes a common clover leaf secondary structure according to proposed model.

(h) Most important variable feature of tRNA is a presence of extra arm based on this two classes of tRNA. The class 1 tRNA has a small arm contains only3-5 bases. Class 2 tRNA has large extra arm consist of 3-21 bases.

(i) tRNA is linked to aminoacid by an ester bond. The secondary structure of tRNA folds into L shape.

(j) The tertiary structure in which the side arm are folded together with the anticodon at one end of the L and acceptor end at the other.

(k) The 'D' and T loops are held by hydrogen bond. All tRNA show common feature of these characteristic folding. However, mitochondrial tRNA exhibit sometime slight different characteristic folding.

(l) The accuracy of protein synthesis depends on the correct aminoacylation of tRNA (Attachment between tRNA and amino acid).

(m) The bond formed between tRNA and amino acid and CCA end is highly energised by storing energy from the ATP which could be utilised in the later stage to form peptide bond at ribosome site (Fig. 8.1).

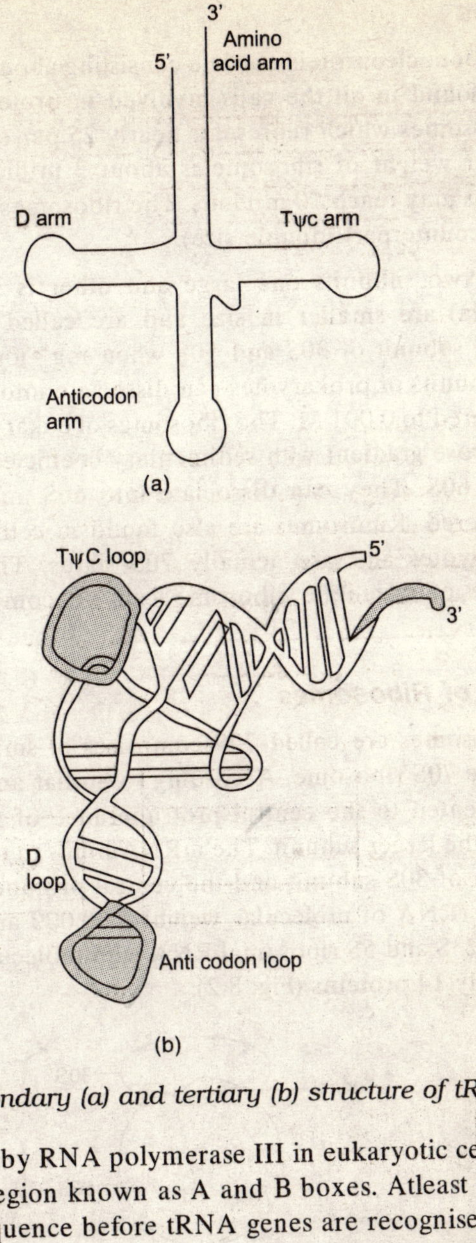

(b)

Fig. 8.1. *Secondary (a) and tertiary (b) structure of tRNA*

(n) tRNA genes are transcribe by RNA polymerase III in eukaryotic cells. tRNA genes have a two separate promoter region known as A and B boxes. Atleast two transcription factors must bind to these sequence before tRNA genes are recognised by RNA pol III.

(o) After transcription of tRNA it undergo some post transcriptional modification. For example, in E. coli tyrosine tRNA containing 3 extra nucleotide after transcription. These extra nucleotide are removed to obtain active tRNA molecules. Other tRNA are synthesised as long precursor that contain 5 to 7 RNA in a single RNA chain. Therefore these nucleotides are cleaved by ribonuclease enzyme during processing.

8.1.3. Ribosomes

It is a granule like ribonucleoprotein particle consisting about 65 per cent RNA and 35 per cent protein. These are found in all the cells involved in protein synthesis. A tiny bacterial cell contains 10,000 ribosomes which represents nearly 25 per cent of the total mass of the E. coli cells. The molecular weight of ribosome is about 3 million. In mammalian cells the total number of ribosomes may reach 10 millions. The ribosomes of eukaryotes are generally larger that its prokaryotic counterpart (double size).

Ribosomes have two subunits one large and other is small subunit. The ribosomes of prokaryotes (bacteria) are smaller in size and are called 70S due to their sedimentation constant. They have subunit of 30S and 50S when mg^{+} ion concentration in the solution is 0.01 M. The 70S subunits of prokaryotes can dissociate into 50S and 30S particle when mg^{2+} concentration is lowered to 0.001 M. The ribosomes of eukaryotes are slightly heavier called 8S and sediment in sucrose gradient with sedimentary coefficient of 80S. The 8S alone have two subunit of 40S and 60S. They can dissociate into 60S and 40S particles, when Mg^{2+} ion concentration is lowered. Ribosomes are also found in cell organelle like mitochondria and chloroplast of eukaryotes and are actually 70S status. They are smaller than eukaryotic ribosomes. The size of organelle ribosome can be comparable to the prokaryotic 70S ribosomes.

8.1.3.1 *Structure of Ribosomes*

The prokaryotic ribosomes are called 70S comprises of smaller 30S and larger 50S subunit associated to form the 70S ribosome. According to Stoffar and Willmann model the long axis of 30S subunit is oriented to the central protruberance of 50S subunit. A tunnel is formed between smaller and the larger subunit. The mRNA binds to the ribosomes and passes between right temporal region of 30S subunit and the central protuberance of the larger subunit. 30S subunit contains 16S rRNA of molecular weight 5,50,000 and nearly 21 proteins. However, 50S subunit contains 23S and 5S ribosomal RNA with molecular weight, 1,100,000 and 40,000 respectively and nearly 14 proteins (Fig. 8.2).

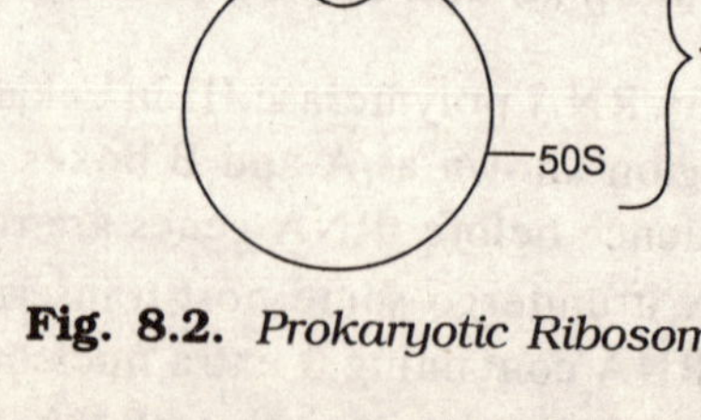

Fig. 8.2. *Prokaryotic Ribosome*

8.1.3.2 *80S Ribosome*

Eukaryotic ribosomes are called 80S comprises of smaller 40S and larger 50S subunit. They are associated with large number of proeins and molecules of RNA. The 40S subunit is curved and ellipsoidal consist of head and a body. It contains molecules of 18S RNA and 30 different proteins (Fig. 10.1) It constitute 45 per cent of the subunit. The larger 60S subunit contains 28S, 5.8 and 5S ribosomal RNA and 40 proteins of molecular weight between 11,500 to 41,000. The RNA constitute nearly 59 per cent of the subunit.

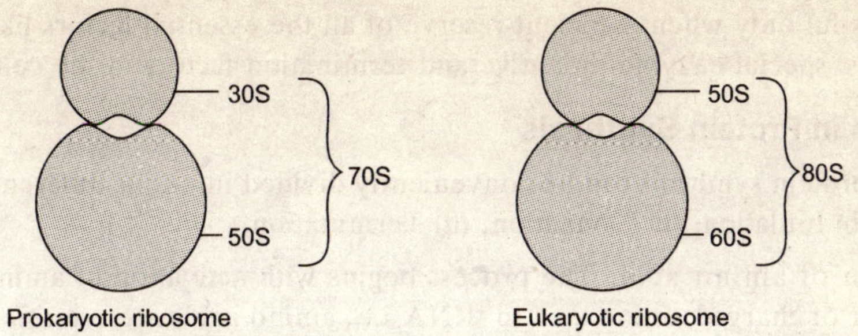

Fig. 8.3. *Ribosomes structure*

8.1.3.3 *Polysomes*

Polysomes are group of ribosomes. These are also called as polyribosomes. Protein synthesis does not takes place on a single ribosomes but on polyribosomes. They may be arranged in a linear fashion or rosette shaped cluster. The number depends on the length of the mRNA molecules.

8.1.3.4 *Functions of Ribosomes*

(a) Take part in protein synthesis.
(b) The ribosome functions as a template, bringing together different component of protein synthesis.
(c) 16S rRNA provides a binding site for tRNA.
(d) Ribosome have a protective function. It prevents mRNA from the enzyme nucleases.
(e) The polypeptide chain when passes between the subunits, they are protected from the action of protein digesting enzyme.

8.2. MECHANISM OF PROTEIN SYNTHESIS

Protein synthesis takes place in all living cell. Synthesis of protein is a highly a coordinated process which involves participation of all the three types of RNA. The three letter genetic code language is transcribed and translated into a product in the cytoplasm. The mRNA contains codons that can base pair with anticodon in aminoacyl tRNA resulted in incorporation of amino acid into a peptide chain. Ribosome acts as a small tiny factories. Association of smaller and larger subunit takes place in the beginning of synthesis and their dissociation takes place at the end of synthesis. Several initiation factors are required for the assembly of ribosomal units. The DNA produce RNA and RNA produce protein. RNA can also produce DNA. This flow of information takes place in all the cells and has been referred as central dogma in molecular biology. Paul za Mecnick in 1950 achieved cell free protein synthesis based on their stuides in the extract of rat liver. They were able to show the essential requirement of energy rich compounds like ATP and GTP for protein synthesis. Careful study reveals that the actual site of proteins synthesis takes place in ribosomes. In the overall process of protein synthesis discovery of tRNA, aminoacyl tRNA synthetase and synthesis of aminoacyl tRNA was independently investigated. The tRNA acts as molecular adapter. It can attached with amino acid. There are 50 different types of tRNA actively participated in protein synthesis. The tRNA adapter for different amino acid varied accordingly in the cell. Protein synthesis would be successful only when sufficient reserve of all the essential factors like three types of RNA, ribosomes, special enzyme, initiative and términation factors in ıhe cell.

8.2.1. Stages in Protein Synthesis

The process of protein synthesis can be conveniently divided into four different stages they are (a) Activation, (b) Initiation, (c) Elongation, (d) Termination.

(a) Activation of amino acid. The process begins with activation of amino acid followed by transfer of charged amino acid to tRNA i.e., amino acids are esterified to their corresponding tRNA with the help of enzyme. Amino acid require aminoacyl tRNA synthetase and ATP for its activation.

(b) Initiation. Involves peptide bond formation between two amino acids. It requires binding of ribosomes to mRNA. The mRNA carries particular codon for particular amino acid and base pair with anticodon present in one of the arm of tRNA.

(c) Elongation. Includes lengthening of peptide chain by continuous addition of amino acids from incoming aminoacyl tRNA. Amino acids are added to the chain one at a time in a step-wise manner. In this process certain proteins known as elongation factors are required for binding of aminoacyl tRNA to the ribosomes. It moves through mRNA chain at the expense of energy rich GTP hydrolysis.

(d) Termination. In this concluding step termination of peptide chain takes place. Synthesis of polypeptide chain stops as soon as ribosome encounter certain signal codon on mRNA. The polypeptide are then released from the ribosome complex with the help of specific releasing factors. In the final stage two ribosomes get dissociated into their free forms. These are then available for next cycle of protein synthesis.

All the above steps are described are overview of protein synthesis. These steps are further described in detail as follows.

8.2.1.1 *I Activation of Amino Acids*

Although more than one hundred different amino acids can be seen in a typical plant cell. Only 20 different amino acids are commonly found in proteins known as protein amino acids. These amino acids are enzymatically activated. Activation of amino acid involves charging of amino acids by ATP. For each 20 different amino acids different aminoacyl tRNA synthetase are available, their function is to charge or activate carboxyl group of amino acids. ATP is utilised in this reaction resulted in the formation of aminoacyl AMP and inorganic phosphate is released. The aminoacyl AMP bind to enzyme complex reacts with amino acid specific tRNA molecule resulted in aminoacyl tRNA (Fig. 8.4).

$$\text{Amino acid} + \text{ATP} \rightleftharpoons \text{Aminoacyl-Amp} + \text{Enzyme} + \text{PPi}$$

$$(\text{Aminoacyl Amp}) + \text{Enzyme} + \text{tRNA} \rightleftharpoons \text{Aminoacy tRNA} + \text{Enzyme}$$

There are more than 50 different types of tRNA are available in a cell. One particular aminoacid are recognised and attached by three to four tRNA. They are called as ISO acceptor'. The enzyme aminoacyl tRNA synthetase are available in all living cells and has a molecular weight of 90,000 dalton. Activity of the enzyme depending on the availability of mg^{2+} ion and the presence of SH group. All these enzyme contain one or two –SH group.

Activation of 20 amino acids (protein amino acids) proceeds in two step. First step is the formation of intermediate aminoacyl adenylic acid. This reaction involves carboxyl group of the amino acid is bound to the 5′ phosphate group of Amp through an anhydride linkage. In the next step aminoacyl group is shifted to free 3′ hydroxyl group of tRNA. The bond between the amino acid and the CCA end of the tRNA is a high energy which stores energy from ATP and this stored energy is subsequently utilised in the ribosome to construct peptide bond.

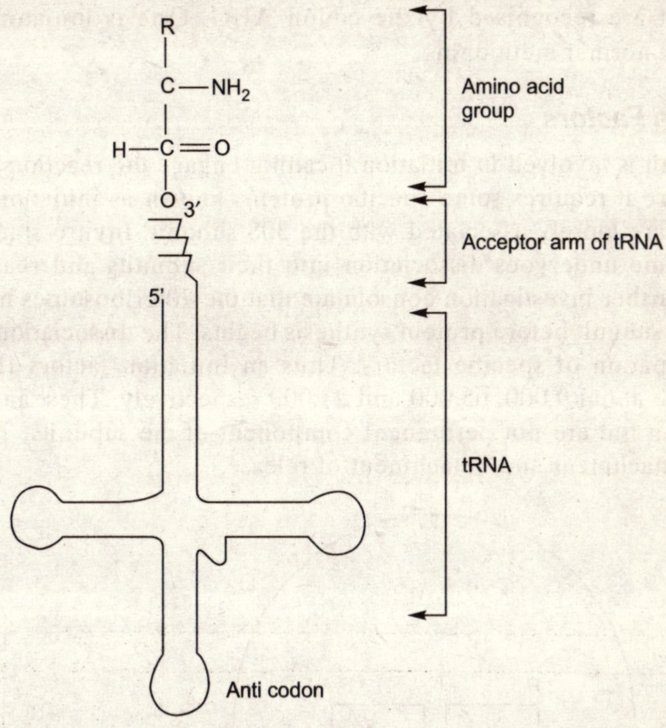

Fig. 8.4. *Activation of amino acid*

8.2.1.2 *Initiation*

The protein synthesis begins with the binding of small ribosomal subunit to the ribosome binding site on the mRNA molecule. The ribosome site contains AUG codons at which protein synthesis starts. This site in prokaryotic mRNA is between three and eight bases. These are complementary to the sequence at 3′ end of the 16S RNA of the small ribosomal unit. The end of 16S RNA in the ribosome is probably hybridises to the mRNA sequence. This region in which aligning the initiating codon in the correct position is known as "Shine Dalgarno" sequence. The correct selection of starting codon AUG is perfectly aided by the secondary structure (hairpin loop) of mRNA.

Methionine is initiating Amino Acid : In bacteria and other prokaryotes synthesis of proteins synthesis starts with the incorporation of methionine as first amino acid to be seen in protein. The initiator codon AUG is present on mRNA represents the methionine amino acid. In almost all bacteria, the starting amino acid is N-formyl methionine (f-met). The amino acid group of the methionine of the methionine group-tRNA is modified by the addition of a formyl group and is described as N-formyl methionyl-tRNA. This tRNA is also known as tRNA f-met. The name of the aminoacyl tRNA is usually referred as fmet-tRNA. The amino acid methionine is modified by charging to become met-tRNA. Then formylation reaction block free NH_2 group (COOH is avaialble to react with NH_2 amino acid during peptide synthesis). This tRNA is used for initiation as it recognises the special initiation codon AUG or GUG on mRNA.

When initiation codon appears at the start of the mRNA f-met is attached. There are two different tRNA that are recognised by the codon AUG. One is initiator f-met tRNA and another is tRNA for normal methionine.

8.2.1.2.1 *Initiation Factors*

Although 30S subunit is involved in initiation it cannot engage the reaction of binding mRNA and tRNA. Therefore it requires some specific proteins known as initiation factor (If). These initiation factor (If) are loosely associated with the 30S subunit. Invitro studies reveals that in bacteria 70S ribosome undergoes dissociation into their subunits and reassociate to initiate protein synthesis. Further investigation consolidate that the 70S ribosomes must first dissociate into its 30S and 50S subunit before protein synthesis begins. The dissociation and reassociation requires the participation of specific factors. Thus an initiation factors IF_1, IF_2 and IF_3 of molecular weight are about 9,000, 65,000 and 21,000 respectively. These initiation factors can attach to 30S subunit but are not permanent component of the subunits. Therefore they can undergo cycles of attachment and detachment of release.

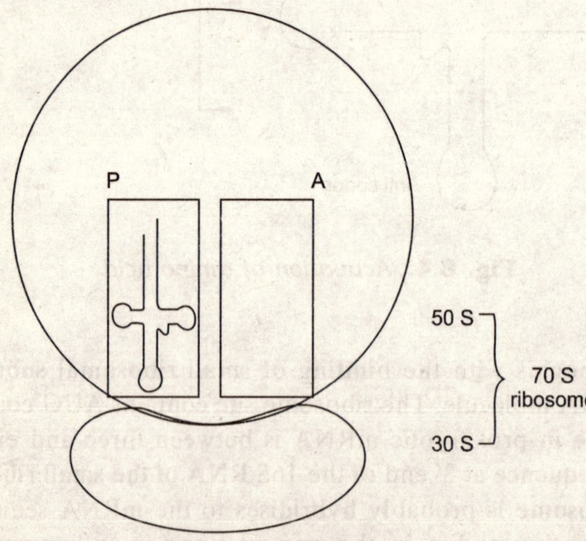

Fig. 8.5. *Initiation complex*

The process begins with the attachment of IF_3 to 30S subunit which then binds to mRNA. IF_3 is also involved in ribosome dissociation once 30S subunit and IF_3 binds to mRNA. IF_1 factor attaches to this complex immediately. It is then followed by attachment of IF_2 with f-met tRNA and GTP and the resulting complex combines with already formed complex of 30S subunit. The systematic joining of all these into a complete complex (IF_1, IF_2, IF_3, mRNA and 30S) is termed as initiation complex. To this entire complex 50S ribosome get attached to form the complete, functional 70S ribosomes. During this process GTP is hydrolysed to GDP and P_1 is released. Once initiation complex is over, initiation factors are then released from the ribosome complex. These released factors can be reutilized by the cells. (Fig. 8.6).

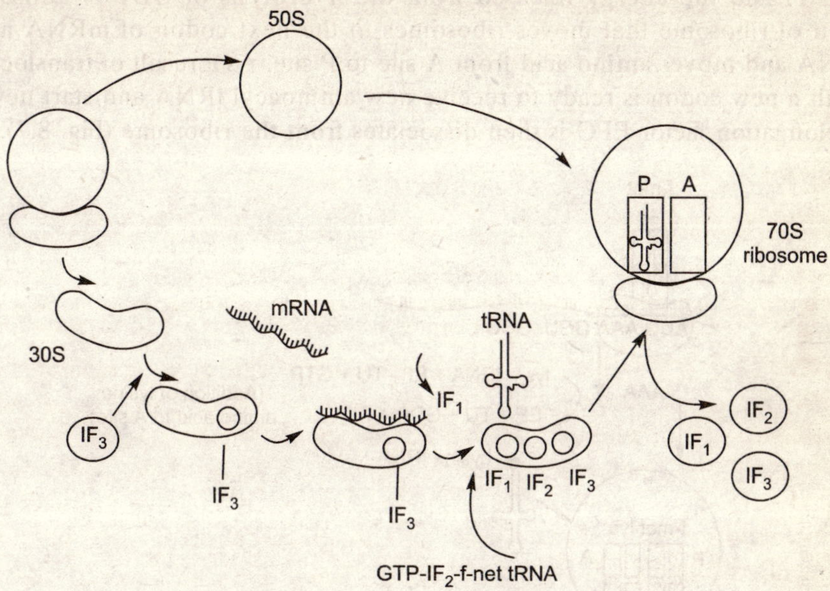

Fig. 8.6. *Organisation of initiation complex*

8.2.1.3 *Elongation*

Elongation process begins once binding of the ribosome subunits (30S, 50S) completes and aminoacyl-tRNA is already in P site and A site is free.

In the complete ribosomal unit 50S subunit contains two binding sites for two tRNA. One is aminoacyl or acceptor site (A site) another is peptidyl or donor site (P site). The peptidyl site is occupied by the presence of f-met-tRNA during initiation complex. Growth of polypeptide during elongation begins with the attachment of aminoacyl-tRNA to the 'A' site of the 70S ribosome. This attachment is mediated by elongation factor (EF-T).

Elongation factor (EF-T) consist of two subunit EF Ts and EF TU. When EF-T combines with GTP it forms EF-TU-GTP complex with the release of EF-Ts. The ternary complex is formed when EF-TU-GTP combines with aminoacyl-tRNA to form EF-TU-GTP aminoacyl-tRNA. This entire ternary complex is then inserted into 'A' site of ribosome. This process is proceeds with the hydrolysis of bound GTP into GDP and inorganic phosphate. The GDP is then leaves the ribosomal site. When f-met tRNA is peptidyl site (P site) and aminoacyl-tRNA on the aminoacyl site (A site). The first peptide bound is formed between the amino acid present in the P and A site. The enzyme peptidyl transferase which is present in the 50S subunit of the ribosome is ready to construct peptide bond by nucleophilic attack of the carboxyl carbon of aminoacyl-tRNA on the esterified carboxyl carbon of the f-met tRNA. The energy is required for this process. In the subseqeunt step there is movement (translocation) of ribosomes along with mRNA in 5′ to 3′ direction and translocation of tRNA from aminoacyl (A site) to peptidyl (P site) site. The newly bound aminoacyl-tRNA on the 'A' site displaces the tRNA from peptidyl-tRNA on the 'P' site in order to form new peptide bond.

During translocation, the movement of ribosomes along with mRNA requires specific protein called elongation factor (EF-G) as well as GTP. The GTP first binds to EF-G and forms a complex EF-G-GTP. This complex is then binds to the ribosome. The GTP is then hydrolysed to GDP and Pi and the energy released from the hydrolysis of GDP is utilised to change conformation of ribosome that moves ribosomes to the next codon of mRNA and displaces peptidyl tRNA and moves amino acid from A site to P site. As a result of translocation 'A' site is vacant with a new codon is ready to receive new aminoacyl tRNA and start new elongation cycle. The elongation factor EFG is then dissociates from the ribosome (Fig. 8.7).

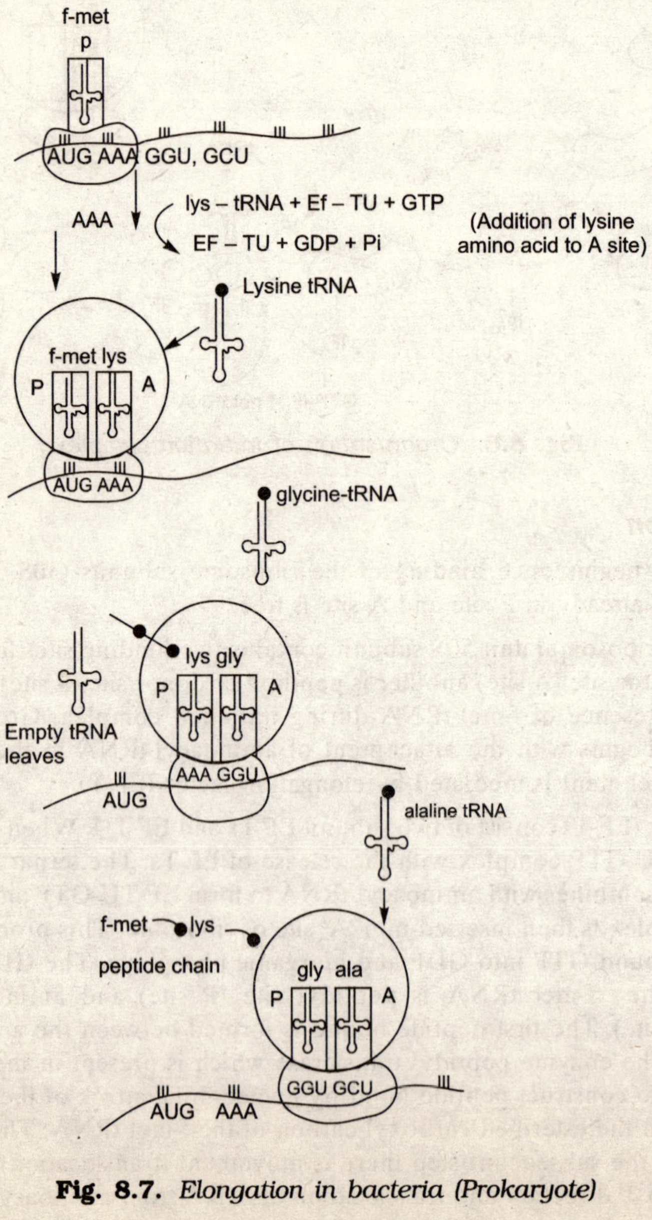

Fig. 8.7. *Elongation in bacteria (Prokaryote)*

8.2.1.4 *Chain Termination*

Termination of polypeptide chain takes place when the ribosomes moves through mRNA reaching termination codon. The termination of polypeptide chain is indicated by one of the three unique termination codon. The chain termination codons are UAA, UGA and UAG. When ribosome reaches one of the special termination codon termination of peptide chain takes place immediately. The release of polypeptidyl tRNA from the ribosome is sponsored by three special proteins called release factors known as R_1, R_2 and R_3. The R_1 is specific for UAG and UAA and R_2 is high affinity for UAA and UGA. When these release factor binds to ribosome which causes displacement of peptidyl tRNA A site to P site. The ester bond between the polypeptide chain is hydrolysed by an enzyme peptidyl transferase. Once the chain is released the last tRNA and mRNA leaves the ribosome and 70S complete ribosome is dissociated into free 30S and 50S ribosome. These free ribosomes will later available for the next round of protein synthesis (Fig. 10.8). The released polypeptide chain in the cell will undergo some modification depending upon the type of polypeptides.

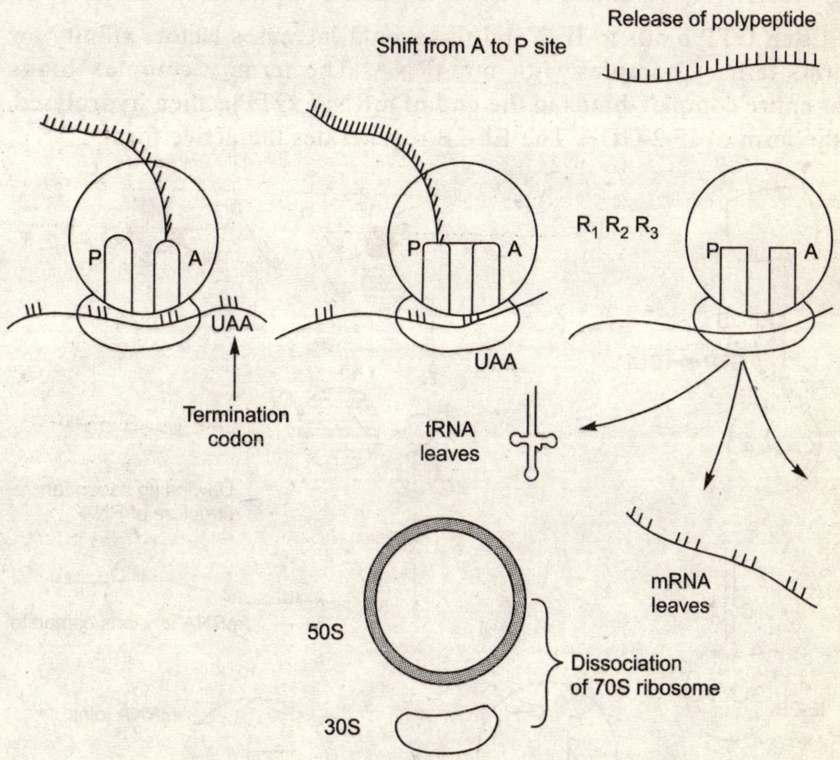

Fig. 8.9. *Termination of protein synthesis.*

8.3 TRANSLATION IN EUKARYOTES

The process of translation in eukaryotes resembles prokaryotic system. However, some order of events are deviated from its prokaryotic counterpart. Following are some of the distinguished features of eukaryotic translation.

(1) The nature of ribosome is different in eukaryotes. Ribosome consist of 40S and 60S together constitute 80S during protein synthesis.

(2) length of the mRNA in eukaryotic cytoplasm is 1000-2000 bases and has methylated cap at the 5′ terminus, carries 100-200 bases of poly (A) tail at the 3′ terminus.

(3) The smaller subunit 40S can able to scan binding site for initiating protein synthesis on mRNA. The binding site for ribosome is initiation codon AUG on the mRNA. Some of mRNA AUG codon is situated 1000 bases distance from starting 5′ terminus. The 40S subunit initially recognises 5′ cap and then migrate along mRNA and stops when it recognises AUG codon.

(4) In eukaryotic cell, the protein synthesis is initiated by special initiator Met-tRNA unlike in prokaryotes methionine is not formylated (methionine is formulated in prokaryote).

(5) As in prokaryotic initial methionine is generally split off from the finished product i.e., polypeptide.

(6) There are more initiation factors. Atleast nine initiation factors have been found in reticulocytes. Eukaryotic initiation proceeds through the formation of ternary complex containing Met-tRNA, IF_2 and GTP. The initiation complex is formed in two steps.

In the first step GTP binds to IF-2 and this would increases factors affinity for met -tRNA. The IF-2 forms ternary complex with met-tRNA. The ternary complex binds to free 40S subunits later entire complex binds to the end of mRNA. GTP is then hydrolysed when IF-2 is released in the form of IF-2-GDP. The EF-2B regenerates the active form.

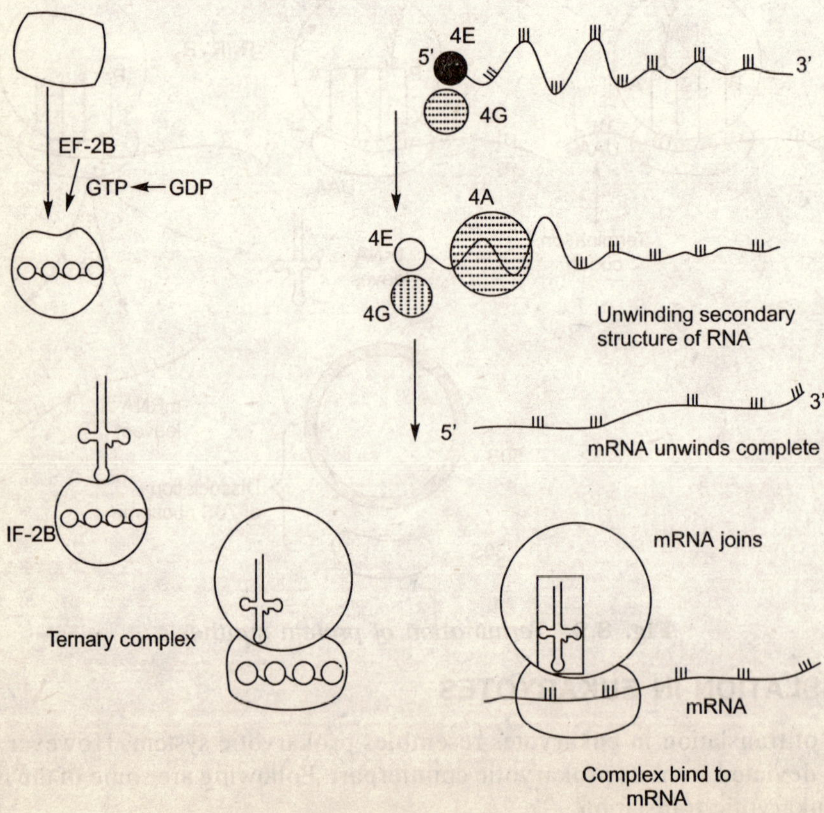

Fig. 8.9. *Eukaryotic Initiation Complex*

Several eukaryotic initiation factors are required to unwind mRNA. Some of the special initiation factors of eukaryotic systems are IF-4F which includes IF-4E, IF-4G, IF-4A, IF-3 and IF-5. Factors like IF-4F/E binds at 5′ cap of mRNA which include IF-4E, 4G, 4A. After IF 4E binds the cap, If 4A unwinds secondary structure that exist in the first 15 bases on the mRNA. The IF 4B also assist further unwinding of mRNA. The energy required for unwinding process is drawn from the hydrolysis of ATP. Recognition of mRNA by ribosome takes place when IF 4G binds to IF-3, which is associated with small ribosomal subunit. The 40S subunit binds to IF-4F and initiate the complex. Factors like IF-3 and EF-6 are required to maintain subunits (40S and 60S) in their dissociated state. IF-3 binds to smaller subunit (40S) and IF-6 is specifically binds to larger subunit (60S). Binding of 60S and 40S subunits cannot takes place until IF-2 and IF-3 have been released from the initiation complex. This is mediated by GTPase (IF-5). All the initiation factors are released when complete ribosomes are formed (Fig. 8.9).

The elongation cycle of eukaryotes is same as in the case of prokaryotes. In eukaryotes however, elongation factor EF-1α is responsible for bringing aminacyl-tRNA to the ribosome. The reaction involves cleavage of GTP to utilise energy.

8.4 POST TRANSLATIONAL MODIFICATION

The production of biologically active functional proteins is very important event in cells once a polypeptide has been made by ribosome. Hence, certain indispensable post-translation modification are required. The primary translation products may need further processing to acquire their full functional properties. Following are some of the important post translation modifications takes place in the cells.

8.4.1 Phosphorylation

Certain amino acids in synthesised proteins undergo phosphorylation. Some of the amino acids undergo phosphorylations are serine, threonine and tyrosine. This process takes place often to control the activity of enzyme. It is generally a reversible process.

8.4.2 Acylation

Some proteins are acylated on the NH_2 terminal α-amino acid group. Such acetyl groups are introduced after completion of protein synthesis.

8.4.3 Glycosylation

Glycoslyation takes place in all eukaryotic cells. A wide variety of chemicals groups can be added to proteins particularly carbohydrate, proteins are added to certain proteins after translocation. Attachment of carbohydrate to proteins are called glycosylation and these proteins are called as glycoproteins. Glycosylation occurs in the golgi apparatus immediately after removal of the leader sequence or peptide sequence. Glycosylation requires certain defined amino acid sequence and are recognised by the certain specific enzymes. The carbohydrate in the glycoprotein perform several functions. They often affect the conformation of the polypeptide. This would strengthen resistant to proteolytic digestion. Hence, presence of carbohydrate can guard protein. In addition, the bound carbohydrate often increase the solubility of protein. The presence of some carbohydrate in protein signifies in classifying proteins to their organelles.

8.4.4 Disulphide Linkage

Certain post translational modifications like disulphide linkage between two cysteine residue present in separate polypeptide chain is one of the most important force in stabilising secondary structure of polypeptide. In protein that contain more than two cysteine residue, formation of disulphide bond is essential for normal structure and enzymatic or hormonal fraction. Disulphide linkage takes place specially in certain membrane proteins and also secretary proteins. In the case of secreted immunoglobulin light chain polypeptide, the disulphide bonds are formed on the lumenal surface of the endoplasmic reticulum (ER). Formation of disulphide bonds helps in stabilising the two separate domains in this protein. Failure to achieve this may result in altered conformation. Another classical example is for disulphide bond formation in the case of insulin. Folding of the nascent proinsulin in the lumen of the ER allows proper disulphide bonds to form formation of disulphide bond between immunoglobulin H and L polypeptide chain occurs in the lumen of the endoplasmic reticulum.

In prokaryotes, N-formyl group of the f-met is removed by the action of enzyme deformylase and the two polypeptide chains are joined by disulphide links.

8.4.5 Protein Targetting

Once proteins are made in the cytoplasm. Many of these proteins are destination to different regions in the cell. The proteins that are destined for export from the cells are synthesised with N-terminal extension of about 15-30 amino acids residues, generally known as leader sequence or signal peptide. These are the special amino acid sequence occupies before the polypeptide chains. In the middle of this sequence is a stretch of amino acid which have hydrophobic side chains which show high affinity for the membranes of endoplasmic reticulum. The function of the signal peptide is to align the growing polypeptide chain, so that it can be drawn through the membrane into a space from which it is ultimately secreted from the cell. Once the leader sequence or peptide signal is through the endoplasmic reticulum it is then removed by hydrolysis with the help of proteolytic enzyme which are generally located in the region. The signal peptide is generally cleaved immediately before amino acid residue with a small side chain.

Several proteolytic cleavage occurs during maturation of secretary and membrane proteins. In some secretary proteins such as lysozyme, growth hormone and certain viral membrane proteins, proteolytic cleavage of signal sequence takes place systematically. After removal they convert directly into the mature active protein. Certain serum proteins such as albumin, insulin and parathyroid hormone are synthesised as long precursors. Hence, proteolytic cleavage is indispensable to obtain active mature molecule.

SUMMARY

There are three types of RNA take part in protein synthesis also known as translation process. mRNA molecule is highly stable in eukaryotic system than in prokaryotes. There are 600 molecules of mRNA can be found in bacterial cell. mRNA contain initiator codon and termination codon. Transport RNA contain four arms and exhibit secondary plus tertiary structure. tRNA genes are transcribed by RNA polymerase III. Ribosomes are tiny cell organelle take part in protein synthesis. It is made up of RNA and proeins.

ribosomes are 70S comprises to 30S and 50S subunits. Eukaryotes contains 40S and 60S together constitute 80S subunits ribosomes.

Translation begins with activation of amino acids by enzymatic process. Activated 20 amino acids is then attached with their respective tRNA. Meanwhile Ribosome (30S) recognise and bind to mRNA chain. The aminoacyl-tRNA is then bind to peptidyl site. It is then covered by 50S ribosomes and meanwhile, several initiator molecule bind to 30S subunit (IF_1, IF_2 and IF_3). At this stage organisation of complete ribosomes, mRNA, tRNA and initiator factors is known as Translation assembly.

Elongation process begins with binding of ribosomal subunits and binding of aminoacyl-tRNA to A or acceptor site. 'P' site already occupied by peptidyl RNA of initiator codon. Attachment of subunits is mediated by elongation factors (EF-T) at the expense of energy. When f-met tRNA in peptidyl site and aminoacyl-tRNA on the aminoacyl site (A site). The first peptide bond is formed between the amino acid present in P and A site. During translocation ribosomes moves alongwith mRNA. There is continuous addition of amino acids into 'A' site followed by peptide bond resulted in peptide chain elongation. Protein synthesis stops when ribosomes arrives to termination codon on mRNA. Protein chain undergo post translational modification like phosphorylation, acylation, glycosylation and disulphide linkage.

QUESTIONS

1. Schematically explain how eukaryotic elongation takes place. (BU May 1996)
2. Explain the process of translation in prokaryotes. (BU April 1998)
3. Explain the structure and role of tRNA. (BU April 1998)
   (BU Oct. 1998)
4. Explain the post-translational modifications in eukaryotes. (BU Oct. 1998; BU Nov. 1999)
5. What are the different types of RNA ? Give one function each. (BU Nov. 1999)
6. Describe the structure and functions of Ribosome. (BU April 2000)
7. Post-translation modifications is essential for protein function and stability. Justify.

SHORT QUESTIONS

1. Structure of Ribosomes
2. Elongation factors
3. mRNA
4. Elongation
5. Termination

(BU April 2000)

9

CHAPTER

Regulation of Gene Expression

Regulation of gene expression is highly a fascinating process seen in all living cells. Tiny microorganisms have their simple regulatory machinery to express their own set of genes. The regulation of specific set of genes is dependent on the nutritional availability in the immediate surrounding. Even simple bacteria make sincere effort to conserve their energy by strictly modulating their gene expression via turned on and turned off mechanism.

Eukaryotic regulation has a definite edge over prokaryotic regulation due to their complexity of genome and gestation period of growth and development. Regulation of gene expression in eukaryotes takes place at various levels. Variety of factors like in-built and external influence play a pivotal role in coordinating differential gene expression.

9.1 PROKARYOTIC GENE REGULATION

Microorganisms are better suited for adaptability to a new environment. Bacteria such as Escherichia coli undergoes series of exposure to variety of new environments in its short span of life. It can exist in the intestinal tract of mammalian system, sewage water, river and lakes. Some of the organisms undergo a wide range of adaptation to extreme cold and hot for survival environments. Thus, it is presumed that natural selection has its own course of selection by imposing adaptation to organism in a new set of environment. Number of microorganisms can reshape their life style in order to sustain in the new environment. E. coli for example exhibit remarkable adaptation potential to wide range of environmental conditions.

When the bacteria enters different environmental conditions, each ecological niche provides different organic molecules as fuel energy molecules. Hence organisms invariably have to utilize these new fuel molecules by reshaping their gene expression. Therefore all kinds of adaptation is due to the ability of bacteria and other organisms to switch on or switch off their specific set of genes. Regulation of specific genes in response to new environmental signal is a fascinating phenomenon. Bacteria turns on its specific set of genes if products are required immediately for their growth in the new environment. They can switch off their set of genes when products are not needed. Thus, by regulating switch on and switch off expression of genes they can avoid wasting their cellular energy, because transcription and translational process utilize considerable amount of energy. Therefore bacteria can divert these conserved energy to synthesize other essential products which can maximize their optimum growth in the environment.

In order to carry out normal cellular activity certain proteins and enzymes are required at all times. Therefore certain genes coded for ribosomal proteins, tRNA and ribosomal RNA are expressed at one time in almost all the cells irrespective of any degree of environmental conditions. Hence, these genes are actively expressed and their products (proteins and enzymes) are required for growth of all cells in all diversified environment. Certain gene products are required only in certain environments and expression of these genes are regulated in such a way that products will be synthesized only when they are required. In this overview of regulation, expression of these genes are continuously being subjected to "turned on" and "turned off" mechanism in response to changes in the environment.

Regulation of gene expression in prokaryotes takes place at different levels like transcription, mRNA processing, translation and enzyme function. Regulation mechanism of rapid turn on and turn off of several set of genes are indispensable for prokaryotes as they are frequently exposed to sudden changes in atmosphere. This nature of regulation of gene expression would help the organism to read just their metabolic process in order to optimize their growth in the new diversified environmental conditions.

9.2 CONSTITUTIVE EXPRESSION

In all the cells certain molecules like tRNA, rRNA and RNA polymerase and certain other enzymes involved in vital metabolic process are absolutely indispensable to carry out basic functions of the cell. Hence, these gene products frequently referred as house keeping functions are essential for all living cells. Genes that are responsible for these products are being constitutively expressed in most cells and are referred as constitutive genes. However, many of the other gene products are required only during certain conditions. Continuous or constitutive synthesis of such products would be a wasteful process as far as energy conservation is considered, because all living cells cannot afford unnecessary expenditure of energy. On the contrary, cells can divert these energy to optimize their growth in diverse climatic conditions. Microorganisms are capable of producing gene products whenever it is required and have definite advantage over other organisms as far as natural selection is considered.

9.2.1 Induction and Repression of Gene Expression

Escherichia coli and other bacteria can grow in presence of any one of the carbohydrate like glucose; galactose and lactose as a carbon source. *E. coli* for example can utilize glucose as a fuel molecule in the environment. If glucose is absent in the environment. *E. coli* cells can grow well on other carbohydrate. When these bacteria (*E. coli*) plated in the medium containing lactose as sole source of energy, *E. coli* cells can intelligently regulate their expression by synthesizing few enzymes to utilize lactose. The enzyme β-galactosidase and β-galactoside lactose permease are essentially required for the lactose catabolisms. B-galactosidase is used to cleave lactose into glucose and galactose and β-galactosidase permease is essential for transporting β-galactosides into the cell. Neither of these enzymes are useful to *E. coli* when present in the environment lacking lactose. The synthesis of these enzymes requires considerable amount of energy in the form of ATP and GTP. Thus *E. coli* cells have articulated a fine regulatory–mechanism by which synthesis of lactose degrading enzymes by turned on in presence of lactose and turned off in its absence. It clearly indicates that *E. coli* cells to not synthesize enzymes most of the times. If cells growing on one of the carbohydrate except lactose is shifted to new medium containing lactose as the only carbon source, they immediately begin to synthesize lactose degrading enzymes by regulating the expression of genes. The genes are turned on in response to substrate in the environment is called induction and the genes whose expressions are regulated are called inducible genes. In addition, if gene products are enzymes, they are called inducible enzymes. The molecules or substance responsible for induction are known inducers.

Bacterial cells exhibit metabolic machinery to synthesize some of the organic molecules like amino acids, vitamins whenever it is required for their growth. For example, E. coli cells have a set of genes responsible for active synthesis of tryptophan. These genes are expressed and synthesize enough amount of tryptophan in absence of it in the medium. When E. coli cells are present in the medium or environment containing sufficient amount or tryptophan for optimal growth their continued synthesis of enzymes for tryptophan biosynthesis is a wasteful process. Their genes are turned off immediately as soon as the cells sense the presence of tryptophan in the medium (or environment). Thus E. coli cells can turn off their expression of tryptophan related genes. A gene whose expression is turned off in this way are called as repressed and when it expression is turned on, it is said to be derepressed.

9.3 THE OPERON CONCEPT

In bacteria, induction and repression of gene expression was first described by F. Jacob and J. Monad in 1961, for which they were awarded nobel prize in 1965. They proposed operon model which explains the regulation of genes encoding the enzymes used for lactose metabolism in E. coli. The word operon is cluster of genes transcribed as a single unit.

In bacteria, gene codes for enzymes are transcribed from adjacent length of DNA with only short-non coding region between them such a super genes are called 'operon' proposed by Jacob and Monad. These cluster of genes are generally transcribed as single units giving rise to polycistronic mRNA. Operon in bacteria act as a single unit which contains structural genes, the operator and the promoter. Much of the information about operon has come from the experiments on mutants. They allowed discovery of two types of genes i.e., structural genes and regulatory genes.

Components of operon : Generally operon consists of structural genes, regulatory genes, the operator and promoter region.

Structural genes : These are single of cluster of genes specifying enzymes which are all part of a single metabolic pathway. They encode information for the synthesis of stable RNA's or mRNA.

Regulatory genes : These genes are not coded by themselves for the production of any functional protein but they give rise to a product involved in the regulation of structural genes. These regulatory genes are situated immediately upstream from the operon. Regulatory genes control the activity of this operon.

Operator : The operator is situated between the promoter and the transcription initiation site. This operator sequence is always located nearest to the structural gene or genes.

Promoter : These are the DNA sequence situated nearer to the structural genes and RNA polymerase binds to this region to initiate transcription. In bacteria, regulatory genes and operator sequence acts as controlling elements to regulate the activity of structural genes. The regulatory genes code for proteins called as repressor. Under certain conditions the repressor protein binds to the operator. When the repressor protein binds to the operator, structural genes will be switched off. Hence, transcription of these structural genes cannot occur. It is because binding of the repressor (protein) to the operator prevents RNA polymerase from binding at the promoter site. Promoter sequence aids in perfect binding of RNA polymerase.

The operon can be studied under inducible operon and repressible operon.

9.3.1 Inducible Operon

Inducible operon is said to be negatively controlled when they are normally not transcribed. The transcription is blocked by repressor binds to operator. The repressor is always a protein. In inducible operon the free repressor binds to operator and sequence on the DNA and switches off transcription of structural genes. When the inducer molecule is present, it binds to

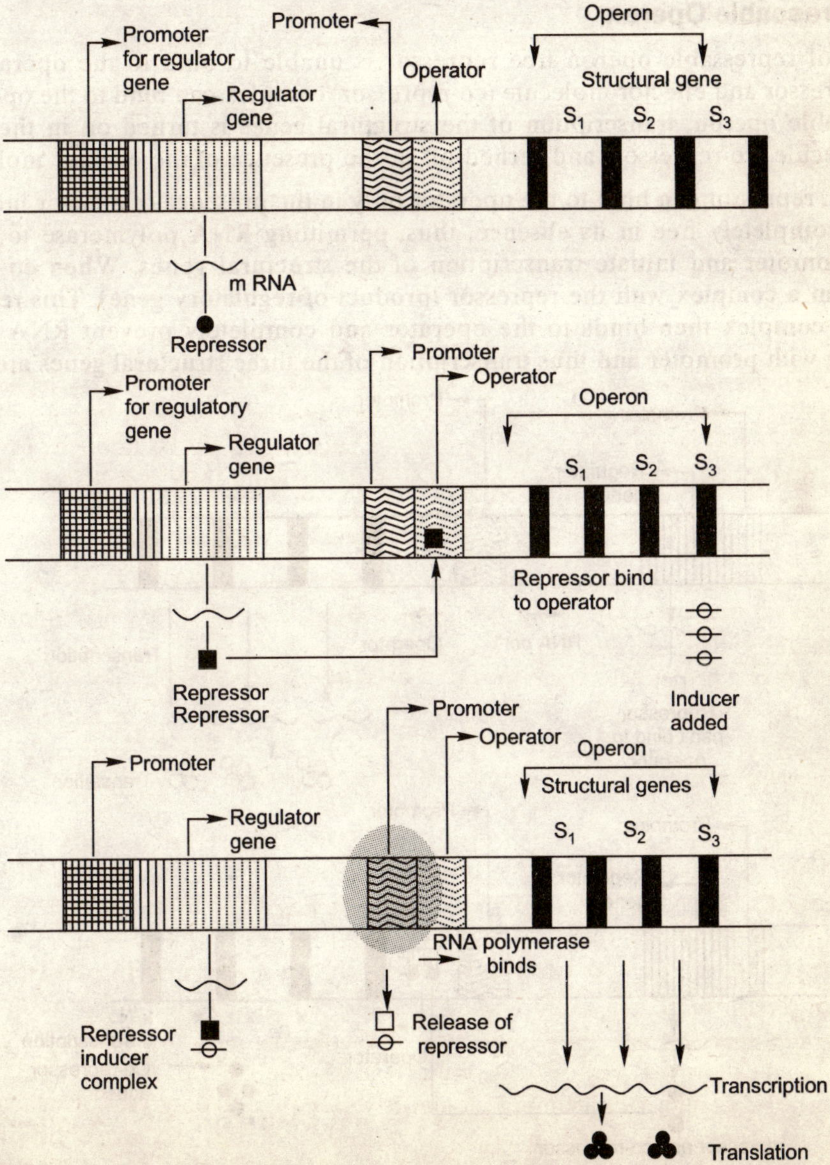

Fig. 9.1. *Induction of operon.*

the repressor (protein) molecule on the DNA. As soon as it binds conformation of repressor molecule is altered and dissociate repressor from the operator. Once the repressor-inducer complex is released it can no longer bind to the operator. Thus, addition of inducer turns on the transcription of structural genes in the operon.

The binding of repressor to the operator regulate transcription of regulator gene is determined by the presence or absence of effector molecule generally amino acids and sugars in the environment the effector molecule. In inducible operon effector molecule could induce operon by binding with repressor. In the case of repressible operon effect (amino acid or sugar) act on repressible operon and are called co-repressors.

9.3.2 Repressable Operon

In the case of repressable operon free repressor is unable to bind to the operator. On the contrary repressor and effector molecule (co-repressor) complex can bind to the operator. Thus in a repressable operon, transcription of the structural genes is turned on in the absence of effector molecule (co-repressor) and turned off in the presence of the effector molecule.

In general, repressor can bind to the operator only in the presence of effector molecule. The operator is completely free in its absence, thus, permitting RNA polymerase to bind to the adjoining promoter and initiate transcription of the structural genes. When co-repressor is added, it form a complex with the repressor (product of regulatory gene). This repressor and co-repressor complex then binds to the operator and completely prevent RNA polymerase from binding with promoter and thus transcription of the three structural genes are prevented.

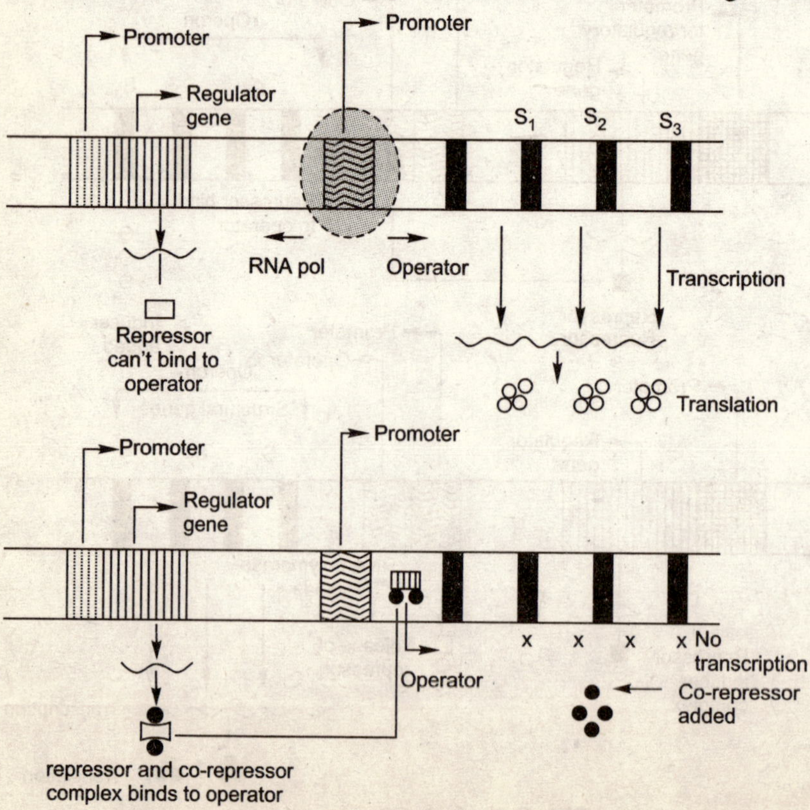

Fig. 9.2. *Repressible operon.*

9.4 THE LAC OPERON

The lac operon is well studied among several other operon. This was the first operon to be discovered and has been intensively studied for the past two decades. Jacob and Monad proposed operon model based on the extensive studies of the lac operon of E. coli. The E. coli bacterium contains about 3000 protein coding genes. Few of them will be expressed at any one time. The classical experiments on lac operon is based on the enzymes involved in the utilization of lactose in the culture medium. As it is negatively regulated, addition of lactose induces the bacteria to synthesise enzymes up to thousand folds in E. coli cells. Regulation of enzymes involved in catabolic degradation of sugar and lipid is found in other bacteria. The substrate molecule acts as a inducer and stimulates the production of enzyme for their degradation.

9.4.1 Components of Lac Operon

The lac operon contains a promoter, an operator, three structural genes and regulator gene. Structural genes are denoted as 'Segment, S_2, and S_3 or z, y, a' and are coding for the enzymes β-galactosidase, β-galactoside permease and transacetylase. These three enzymes are required for the utilization of lactose. Structural genes (z, y and a) in the operon can be controlled by turned on and off in a unified manner. The genes for the three enzymes are transcribed together by RNA polymerase and are transcribed as a single polycistronic lac mRNA. The function of β-galactoside permease is to pump lactose inside the cell. Whereas β-galactosidase cleaves lactose into glucose and galactose. The function of the other enzyme transacetylase is not known.

Expression of lac operon is regulated by lac repressor. It is basically a protein. The lac regulator gene designated as i gene codes for the lac repressor protein having four subunits containing 370 aminoacids. The repressor binds strongly to the operator sequence which is located close the β-galactosidase gene, preventing RNA polymerase from binding to the promoter and transcribe structural genes.

The binding activity of repressor takes place only if inducer is absent. The actual inducer of lac operon is 'allolactose'. Its synthesis from lactose is catalysed by the first enzyme β-galactosidase. The repressor contains four subunits, each subunits has one binding site for the inducer (lactose). Repressor undergoes conformation change after lactose binds to it. As a result it is unable to bind to the operator region and transcription of z, y and a genes takes place immediately.

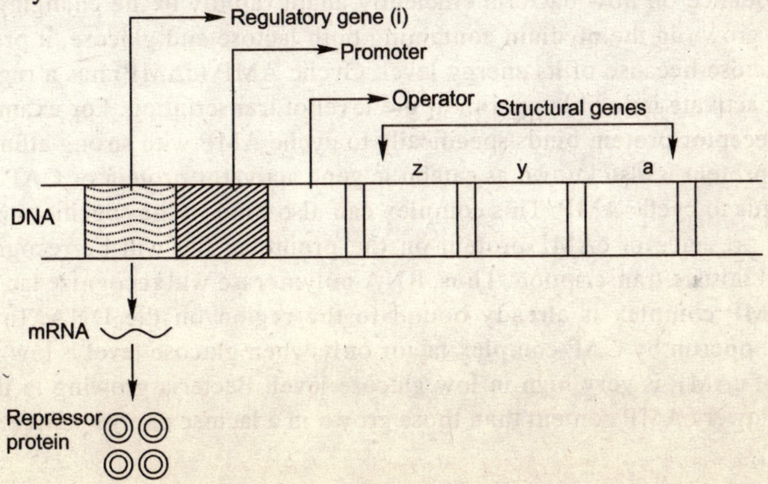

Fig. 9.3. *Structure of lac operon.*

E. coli is adapted to rapidly changing nutritional environment. While in the absence of lactose only 3 molecules of β-galactosidase are synthesized in uninduced state, providing a low background level of enzyme activity when lactose (inducer) is added.

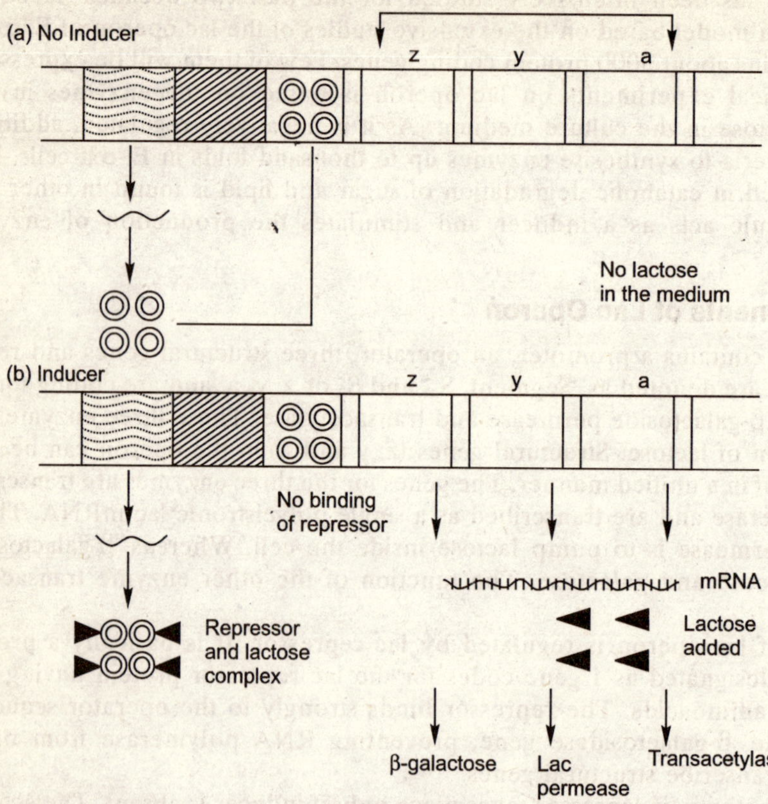

Fig. 9.4. *(a) and (b) Lac operon in non induced and induced state.*

9.5 POSITIVE REGULATION OF LAC OPERON (CATABOLIC REPRESSOR)

Lac operon is also under positively regulated. Regulation of lac operon by positive regulation is a classical evidence on how bacteria efficiently adapt rapidly to the changing environment. When E. coli is grown in the medium containing both lactose and glucose, it prefers to choose glucose than lactose because of its energy level. Cyclic AMP (cAMP) has a regulatory role in bacteria. It can activate inducible operon at the level of transcription. For example, E. coli has a cyclic AMP receptor protein binds specifically to cyclic AMP with strong affinity. This cyclic AMP receptor protein is also known as catabolic gene activator protein or CAP. It is a dimeric protein and binds to cyclic AMP. This complex can also bind to the specific site within the lac promoter. The presence of cAMP-protein on the promoter site will be recognized by RNA polymerase and initiate transcription. Thus, RNA polymerase will recognize lac promoter only if the CAP-cAMP complex is already bound to the region on the DNA. Interestingly the induction of lac operon by CAP-complex factor only when glucose level is low in the medium because level of cAMP is very high in low glucose level. Bacteria growing in the presence of glucose have a lower cAMP content than those grown in a lactose energy source. When cellular cAMP level is low due to high glucose, the CAP protein do not bind to the promoter and the lac operon is repressed even in the presence of lactose. Hence, bacteria utilize glucose efficiently than lactose due to its rapid adaptation to the changes in natural environment.

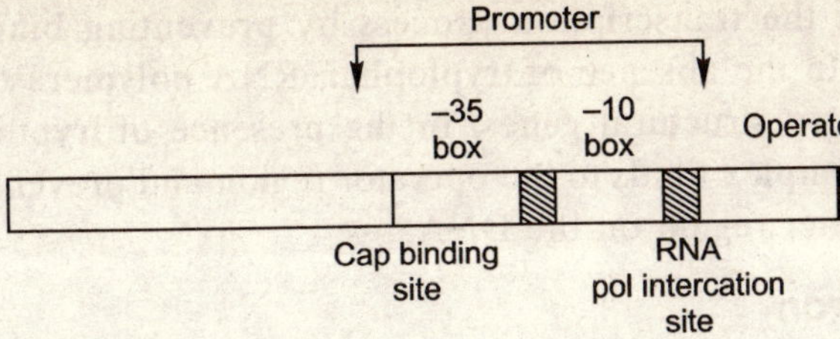

9.6 THE TRP OPERON

Regulation of enzyme repression is found in several bacteria involved in the synthesis of amino acids or nucleic acid precursors. There are number of operons containing the structural genes coding for the enzymes required for the synthesis of amino acids. These structural genes in the operon are repressed, either by the amino acids themselves so that unnecessary and wasteful synthesis of these amino acids does not occur when there is sufficient amount of amino acids available to the cell.

The trp operon consists of five structural genes required for the synthesis of tryptophan. The expression of this operon is regulated by the availability of tryptophan in the culture media and the presence of tryptophan repressed the synthesis of trp enzymes. Five trp structural genes are trpE, trpD, trpC, trpB and trpA which code for enzymes such as Anthranilate synthetase, phosphoribosyl anthranilate transferase, phosphoribosyl anthranilate isomerase, tryptophan synthase β and tryptophan synthase α respectively.

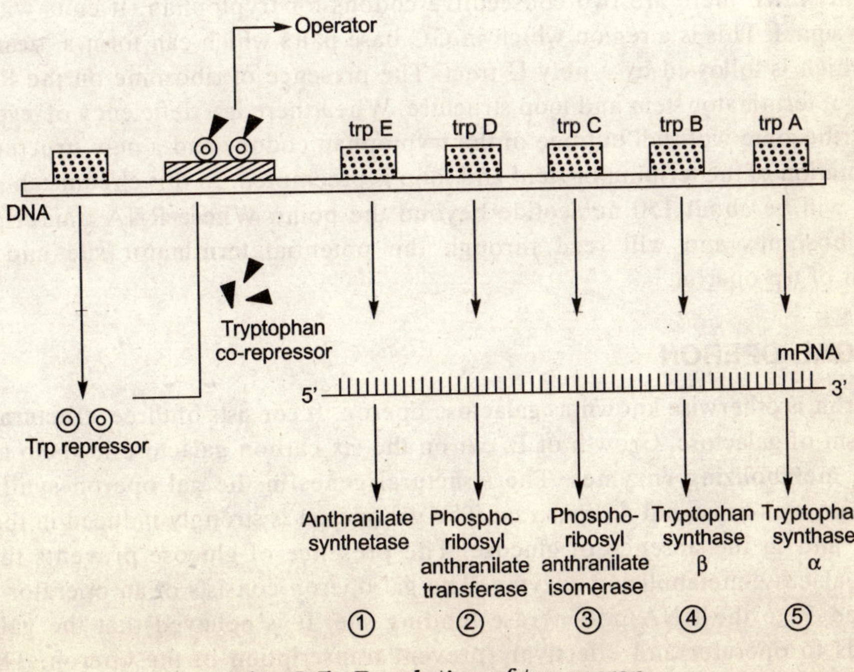

Fig. 9.5. *Regulation of trp operon.*

In enzyme repression in trp operon the regulatory gene produces a repressor protein which is normally in an inactive state. When the repressor-protein combines with co-repressor (here the co-repressor is amino acid), it undergoes a conformation change, which enables to bind to the operator region and inhibits the transcription process by preventing binding of RNA polymerase to the trp promoter. In the absence of tryptophan RNA polymerase binds to the promoter and transcribe all five trp structural genes. In the presence of tryptophan the co-repressor and repressor protein complex binds to the operator region and prevents the binding of RNA polymerase to the promoter region on the DNA.

9.6.1 Control of the trp Operon

As we have seen earlier in trp operon regulation, the trp operon is also subjected for second level of regulation which takes place at the level of termination of RNA synthesis. In mutant cell, the trp R mutants which fails to make repressor, addition of tryptophan into the media devoid of tryptophan will cause 10 fold decrease in the rate of synthesis of the tryptophan biosynthesis enzymes. This second level of regulation of the trp operon is called attenuation and the sequence within trpL that control regulation is called the attenuator. Normally 90% of the RNA polymerase molecule that starts transcription at the promoter region terminate at attenuator site which result in the synthesis of short RNA less than 10 % of the molecule continue to produce trp mRNA. The level of tryptophan in the cell can control termination process at attenuator site, when the tryptophan concentration is low, full length mRNA proportion is increased drastically. When it is high most of the mRNA terminate prematurely. This results in premature termination of trp operon, transcription occurs only in the presence of tryptophan yields of leader-sequence transcript.

The RNA transcription from the leader sequence contains a short open reading frame of 42 base pairs, in which there are two consecutive codons for tryptophan. It ends with a typical termination signal. This is a region which in GC base pairs which can form a 'stem' and 'loop' structure, which is followed by a poly U tract. The presence of ribosome on the RNA allows termination of terminator stem and loop structure. When there is a deficiency of tryptophan, in the cell the ribosome will stall out one of the tryptophan codons and a new structure arises so that the formation of the terminator stem and loop is preempted. In this circumstance the RNA polymerase will be about 150 nucleotide beyond the point. Where RNA transcript emerges from the ribosomes and will read through the potential terminator site and continues transcription of trp operon.

9.7 THE GAL OPERON

The gal operon is otherwise known as galactose operon. It consists of three structural genes for the catabolism of galactose. Growth of E. coli on the six carbon galactose leads to an increase of galactose metabolizing enzymes. The structural genes in the gal operon synthesizes the enzymes kinase, transferase and an epimerase. The gal operon is strongly induced in the presence of galactose and in the absence of glucose. The presence of glucose prevents the induced increase of galactose-metabolizing enzyme. The gal operon consists of an operator site which was identified 5′ to the RNA polymerase binding site. It is believed that the gal repressor protein binds to operator and effectively prevent transcription of the operon. This operon consists of two promoter sites separated by 5 thinth bp. The pribnow boxes overlap and are used to direct the RNA polymerase to initiate transcription. Binding site at 3′ promoter site is known as 'S' and transcription is stimulated by CAP binding at – 35 box. There is a second binding site S_2 for CAP at the 5′ side. The CAP binding inhibits transcription from the other promoter site which is used only in the absence of CAP. Recent studies reveal that the gal operon contains two initiation sites for RNA synthesis. One site is S_2 which binds the polymerase and initiate transcription of genes for galactose metabolizing enzymes at low level of cyclic AMP or CAP. The second binding site is 'S' and it requires high level cyclic AMP and CAP in order to facilitate RNA polymerase to bind and transcribe all the three galactoside enzyme. It seems that the S_1 polymerase binding site exhibit higher affinity for RNA polymerase than does S_2, when cyclic AMP is high and the rate of RNA chain initiation is greater and more galactose mRNA is produced.

To summarise, gal operon has two transcription sites one binding site S_2 is active continuously even in the presence of glucose. So it does not require CAP or cyclic AMP the second binding site is S_1, and requires galactose, CAP and high level of cAMP. High level of cAMP is brought about by the absence of glucose on full induction of the gal operon in the presence of CAP-cAMP. Production of kinase level is substantially higher than the other two enzyme i.e., epimerase and transferase.

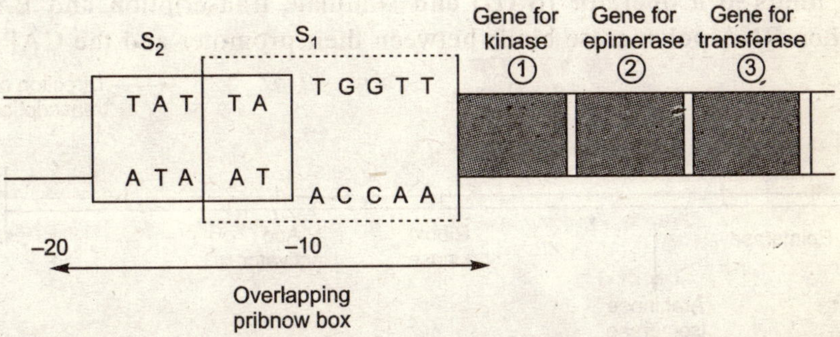

Fig. 9.6. *Regulation of gal operon.*

9.8 THE HUT OPERON (*HISTIDINE*)

The hut operon (histidine utilization) has been most studied in Salmonella typii. The two species of bacteria (Klebsiella and Salmonella typhimurium) can degrade amino acid histidine into glutamic acid, ammonia and formamide. These products provide carbon and nitrogen source. Two groups of hut enzymes are required for histidine degradation. The hut operon is switched on when there is plentiful supply of amino acid histidine in the medium. Excess histidine can be degraded by bacteria.

There are two separate operons present for the synthesis of hut enzymes. In the absence of histidine each operon has its own operator and promoter. The four genes required for the catabolism are organized in two operon, which are adjacent. In the absence of histidine, transcription of both the operons are switched off by single repressor protein. The two identical repressor proteins bind at two different operator sites and block transcription. When histidine is available it acts as an inducer by combining with repressor molecules which alter the configuration of the repressor. Altered repressor cannot bind to the operon and transcription takes place immediately. The repressor of hut operon is added by a gene in one of the two operons and the synthesis of hut repressor is under the control of inducer molecule i.e., histidine. When the level of inducer is reduced repressor molecule accumulate which shut down the synthesis of more repressor. Catabolism repression is also involved in the control of these operons since CAP-cAMP stimulates transcription of one of the right handed operon.

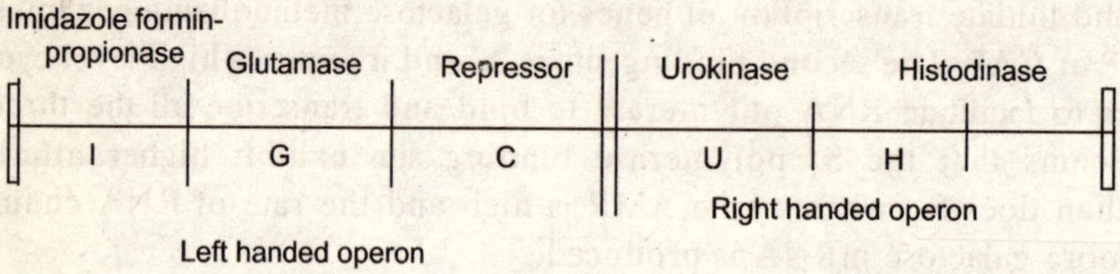

The ara operon : The ara operon comprises of three structural genes known as ara B, A and D (BAD) codes for the synthesis of catabolic enzymes. The sugar arabinose is degraded by these ara enzymes. In the operon gene for repressor (ara C) is situated very close to structural genes. It is separated from them by only 140 bp. Its transcription is in the opposite strand of DNA. The ara repressor protein binds to the operator and represses its own transcription. This is a classical example of auto regulation. When arabinose is added, it combines with repressor and complex binds to a operator (BAD) and stimulate transcription and BAD genes are transcribed when RNA polymerase binds between their promoter and the CAP-binding site.

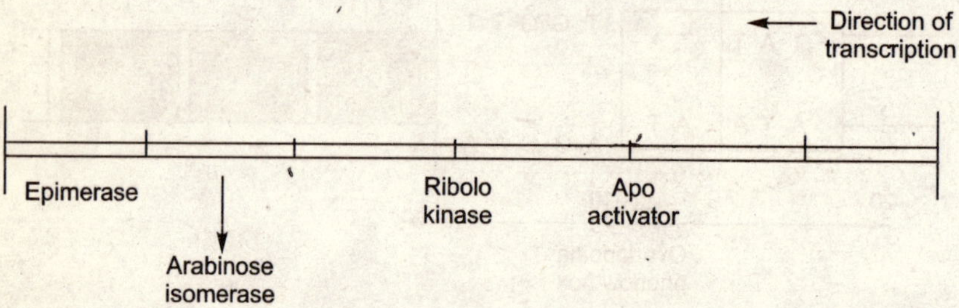

Fig. 9.7. *The ara operon.*

9.9 GENE REGULATION IN EUKARYOTES

Eukaryotic organisms comprise a great deal of diversity in their cell types. Two hundred cell types have been identified in human beings. Various cell types are bone marrow (B), Thymus (T), lymphocytes, lung epithelium, liver, muscle and cell nerve. Although they differ in their morphological feature the genetic information of various cells in an organism is virtually the same. The differential nature of cell types seems to be born mainly due to the expression of specific types of proteins and mRNA. However, exceptions are the set of genes which are constitutively expressed in all the cell types due to their house keeping functions.

Eukaryotic system is a complex organisms. They have evolved with a complex array of regulatory mechanisms in which only a specific set of genes are expressed in one cell type but not in other. Without compromising their sequence of gene itself. The expression of specific set of genes involved in photosynthesis is confined to leaf cell types, but not in root cell types. In addition, the growth and development of an organism a specific group of genes are expressed during the onset of reproduction process. In principle, gene expression can be regulated at various levels during the process of DNA to RNA and RNA protein. Hence regulation mechanisms are active during the process of transcription, RNA processing, RNA modification, mRNA degradation, translation and post-translational modification of protein activity. Several models and regulatory mechanisms have been postulated among eukaryotes.

9.9.1 Davidson and Britton Model

Regulation of gene expression in eukaryotic system was explained by R.J. Britton and E.H. Davidson. This model virtually deviates from the operon model which explains prokaryotic gene regulation, in which genes are clustered or adjacent. In eukaryotes the regulatory genes are distributed in different chromosomes. The genome of eukaryotic system is large and most of the genes are in repressed state. However, large part of the genome are permanently repressed.

According to this model eukaryotic genome consists of large number of sensor sites and integrator genes are situated adjacent to sensor sites. The structural genes are located in various regions of the different chromosomes and they are in repressed state. The sensor site can recognize various signaling agent including substrate of specific enzymes or hormone receptors. When a specific substrate binds the sensor, it becomes activated and initiate transcription of the adjacent integrator gene. This integrator gene produce activator RNA. Once activator RNA is produced it can combine with receptor molecule present within the same chromosome or in different chromosome. Subsequently structural genes are transcribed by RNA polymerase to yield mRNA, which are then translated to produce protein.

9.10 GALACTOSE REGULATION IN YEAST

Yeast (Saccharomyces cerevisiae) contains genes which code for regulatory proteins for a number of metabolic pathways. This group of regulator proteins act as both positive as well as negative regulatory proteins. Regulation of galactose utilization has been studied extensively in yeast cells. Several regulatory genes are present which codes for three enzymes devoted to convert galactose into glucose phosphate. The three enzymes are galactokinase, coded by a gene GAL-1, galactose transferase coded by GAL-7 and epimerase coded by GAL-10. The three genes (GAL 7, 10 and 1) for the enzymes are located on chromosome II of yeast cells. In addition, a fourth gene which is meant for galactose transportation is located on Chrome XII. A fifth gene is present on another chromosome encodes a-galactosidase which convert melibiose to galactose in the beginning of the process. Galactose acts as inducer in expressing all the five genes through a fine mediator i.e., gal 4 protein. All the five genes are never expressed unless gal 4 protein is present in a cell. Hence, gal 4 is a key player in gene regulation and play a significant role in the utilization of galactose in yeast cells.

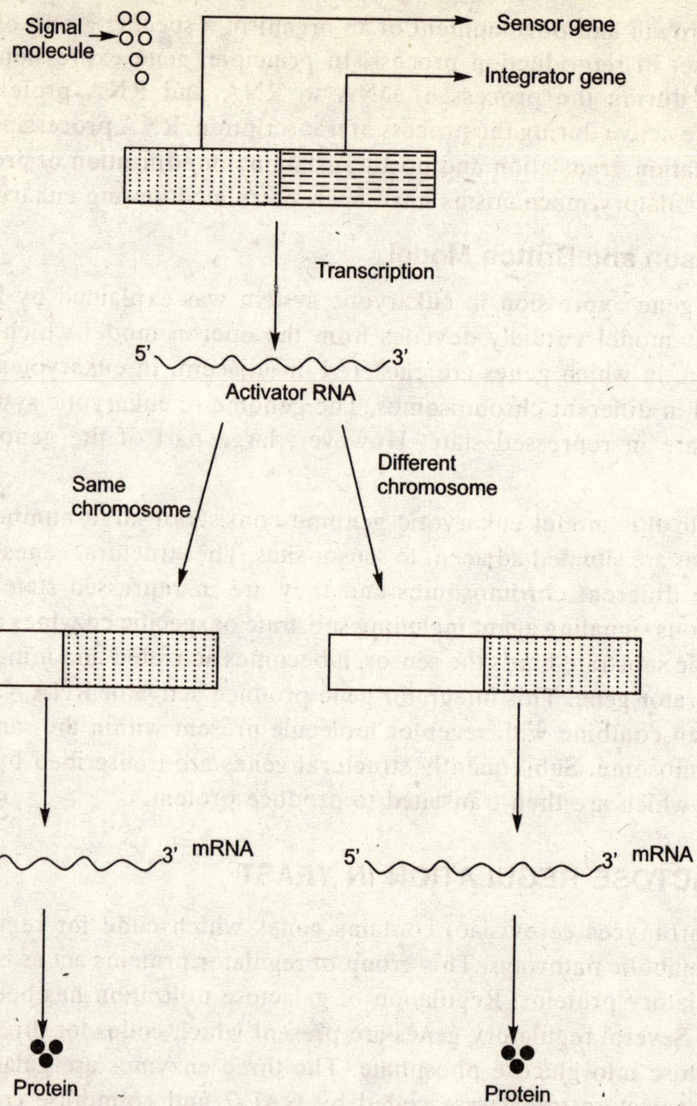

Fig. 9.8. *Eukaryotic gene regulation model by Davidson and Britton.*

The GAL-4 protein acts as in positive regulation. It acts by binding to specific DNA sequence upstream from the site initiation and transcription of GAL-1, GAL-7 and GAL-10 gene. In the presence of galactose, genes for galactose utilizing enzyme is turned on and switching off transcription of three GAL genes in its absence.

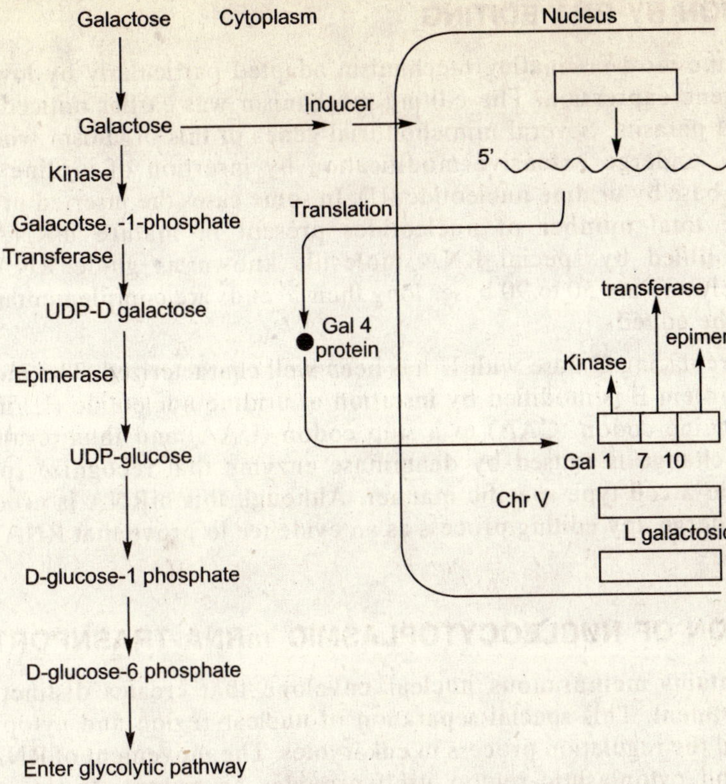

Fig. 9.9. *Regulation of gal operon in yeast.*

9.11 REGULATION IN NEUROSPORA

Neurospora cells contain enzymes for the synthesis of many aromatic amino acids like tyrosine and phenylalanine. When the level of these aromatic amino acid increases beyond its cell requirement, certain scavenger enzymes are synthesized by quicnoic acid as an inducer. High level of this acid acts as an inducer in expressing scavenger enzyme, which in turn degrade high level aromatic acids. Once quicnoic acid is degraded expression of scavenger enzyme is turned off. The genes coded for scavenger enzyme are dehydrogenase, dehydroquinase and dehydralase. The genes for the above enzymes are qa-3, qa-4, qa-2 are present in a cluster and expression system are under positive control. The inducer quinic acid seems to bind with a regulator, products of a gene qa-1, to turn on the expression of the qa genes.

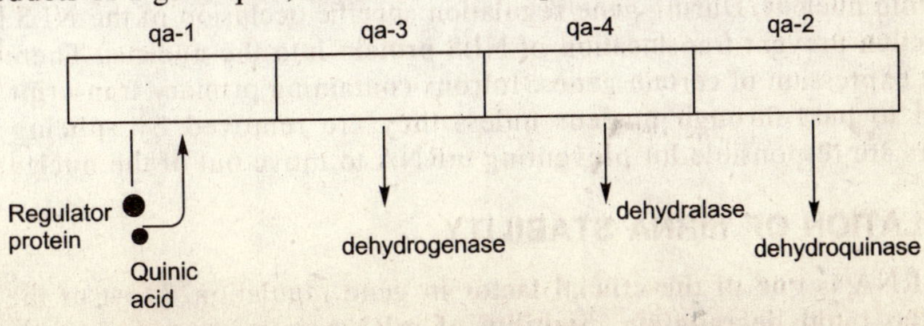

Fig. 9.10. *Regulation of Neurospora by quagene cluster.*

9.12 REGULATION BY RNA EDITING

Editing of RNA is the most fascinating mechanism adapted particularly by lower eukaryotes for the control of gene expression. The editing mechanism was earlier noticed in unicellular trypanosoma, blood parasite. Several mitochondrial genes of this organism when transcribed to produce mRNA, undergo extensive modification by insertion of uridine nucleotide or replacement of one base by uridine nucleotide (U). In some cases the inserted uridine occupies almost half of the total number of nucleotides present in mature mRNA. The initial transcription is modified by special RNA molecule known as guide RNAs. These are transcribed separately and are 50 to 90 bases long their 5′ ends are complementary to sequence in the transcript to be edited.

RNA editing by replacing C base with U has been well characterized. The messenger RNA of intestinal apoloprotein B is modified by insertion of uridine nucleotide (U) in place of 'C'. This converts glutamine codon (CAA) to a stop codon (UAA) and thus result in truncated protein. This base change is caused by deaminase enzyme that recognize some structural features of mRNA in a cell type specific manner. Although this mRNA is produced in liver cells they do not undergo any editing process as an evidence to prove that RNA editing is cell type specific.

9.13 REGULATION OF NUCLEOCYTOPLASMIC mRNA TRASNPORT

Eukaryotic cell contains membranous nuclear envelope that creates distinct nuclear and cytoplasmic compartment. This special separation of nuclear region and cytoplasmic region has greatly enhanced the regulation process in eukaryotes. The movement of RNA and protein between nucleus and cytoplasmic region itself provides an unique opportunity for gene regulation. Several thousands of nuclear pore complexes are distributed over nuclear envelope and these are crucial for the translocation of various types of molecules in both the directions. It consist of large number of proteins and it can allow free diffusion of molecules upto 20,000 daltons. In a situation larger molecules can be translocated they can move by energy dependent active process, probably by receptor mediated transport. Several proteins and RNA molecules move in one direction (unidirection) or can translocate between nucleus and cytoplasmic region.

Gene regulation can be seen at the level of translocation of some specific proteins. Certain proteins actively travel from the cytoplasm to the nucleus due to the presence of signal peptide generally called as nuclear translocation (NLS). This signal peptide contains short lysine and argenine rich peptide sequence is believed to be recognized by cytoplasmic based protein factors in the specific interaction with a component of the nuclear pore complex and leading to translocation into nucleus. During gene regulation specific occlusion of the NLS by a protein-protein interaction prevent translocation of NLS protein into the nucleus. Thereby indirectly retard specific expression of certain genes. Introns containing primary transcript (mRNA) are never allowed to pass through nucleus unless they are removed by splicing mechanism. Splicing factors are responsible for preventing mRNA to move out of the nucleus.

9.14 REGULATION OF mRNA STABILITY

Stability of mRNA is one of the crucial factor in gene regulation. Most of the prokaryotic mRNA undergo rapid degradation. Stability of mRNA is in response to changes in the environment by expressing other set of genes. Hence, prokaryotic mRNA are less stable in the cell. Eukaryotic system on the contrary exhibit longer life of mRNA and are potentially stable than prokaryotic mRNA. The β-globin mRNA is stable for more than 10 hours. Some of the mRNA codes for protein acts as a growth factor have a shorter life span with a half life less than 30 minutes. Many short lived in mRNA have shown to contain conserved AV rich sequence of about 50 nucleotides.

Another classical example is the stability of mRNA for histone protein. Histone mRNA stability is regulated at 'S' phase of cell cycle. At the synthesis (S) phase of cell cycle as more amount of DNA is synthesized, there is a considerable increase in the demand for histone proteins. Degradation of histone mRNA takes place immediately once 'S' phase of the cell cycle is completed. The regulation of histone mRNA stability is due to a short 3′ stem-loop structure and replaces poly (A) tail found in other mRNA.

9.15 REGULATION BY DNA METHYLATION

Most of the higher plants and animal DNA is often modified by addition of methyl group to one of its base. The cytosine base is enzymatically modified to 5′-methyl cytosine. This methylation process is known as DNA methylation. Methyl group in the 5′ carbon of pyrimidine occupy exposed regions within the major grooves of DNA molecules. Thus, they plan an influential role in the interaction of DNA with specific proteins.

There is a strong evidence to prove that DNA methylation is linked with gene expression. Information on the involvement of methylation in the control of gene expression in eukaryotes is based primarily on three evidences : (1) various studies have demonstrated that a correlation between the level of gene expression and the degree of methylation i.e., low methylation is correlated to high level of gene expression. Similarly high methylation can be correlated to low gene expression ; (2) methylation pattern is tissue specific in some cases and (3) Azacytidine is a drug which inhibits methylation process after its incorporation into DNA which results in increased expression of genes. Over 90 per cent of DNA methylation in the DNA of most eukaryotes occur in CG dinucletoide sequence.

In vertebrates, cytosine methylation takes place in more orderly fashion since the same DNA sequences are methylated in one cell type but not in others. Prominent example is β-globin gene. It is methylated in non producer cell, but unmethylated in erythroid cell when it is transcribed. All house keeping genes are unmethylated even though they are rich in CG dinucleotides. This is probably due to their commitment to house keeping function. These unmethylated CG rich sequence contains enhancer or promoter sequence called C_pG islands. In addition, differential methylation of single gene or entire chromosome region can occur in a special cell type. Specific example is β-globin or in a specific chromosome (X chromosome).

In the methylation process site of methylation is more crucial. Frequent methylation at promoter region can inhibit expression of genes by blocking the binding of transcription activator proteins. However, extensive methylation at coding region of gene may have no effect on gene expression.

9.15.1 DNA Hypomethylation

The chromatin regions contains actively transcribed genes. These regions are hypersensitive sites and represent binding sites for regulatory proteins in enhancer. These sites are present within or adjacent to genes. Binding of transcription factors to these regulatory regions prevents tight chromatin packaging and allow easy expression of genes. Hypomethylation is virtually free from methylation process and are cell specific. Prominent example is hemoglobin gene has hypersensitive sites in erythroid cells but not in the cells of liver or brain. The hypersensitive sites are formed when transcription factors specifically bind to these sites at a particular stage of differentiation. These results provide an evidence to the correlation of gene transcription to decondensed chromatin and DNA hypomethylation.

9.16 TRANSCRIPTION FACTORS

Eukaryotic transcription is a highly coordinated process which involves participation of the enzyme RNA polymerase and proteins. RNA polymerase enjoys all privileges in prokaryotic transcription as they do not require the support of any proteins. However, the situation is different in a eukaryotic system. RNA polymerase alone cannot perform the transcription. Hence, certain group of proteins known as transcription factors are indispensable to initiate the process. RNA polymerase plus transcription factors together called basal transcription apparatus forms a stable complex at transcription initiation site by binding to TATA region in the promoter sequence of DNA.

Transcription factors are generally divided into general transcription factors (GTF) and sequence specific transcription factors (SSTF). General transcription factors are indispensable to transcribe all RNA pol II dependent genes when sequence specific transcription factors regulate transcription. There are six general transcription factors (GTF) are required in order to transcribe RNA polymerase dependent genes. These are TF II A, TF II B, TF II D, TF II E, F and H.

9.16.1 Transcription Factors for RNA Polymerase II

Transcription factors for RNA polymerase II (TF II A, TF II B, TF II D, TF II E, F and H) are basically DNA binding protein and protein interactive polypeptides. TF II D transcription factor can recognize the promoter that contains TATA sequence and binds specifically at this region. TF II D contains two types of components TATA binding protein (TBP) and another subunit TBP associated protein or factors (TAF). TBP is a small protein of 30 kD which recognizes TATA box in the promoter. TF II D contains various TAFs that can recognize different promoters. The total mass of TF II D is 800 kD. The mass of the components of TF II D i.e., TBP and TAF varies from 40-240 kD. TBP is associated with TAF and binds to the DNA at promoter region. TBP has a unique property of binding to the minor groove of DNA. The inner surface of TBP binds to DNA whereas its large outer surface makes contact with other proteins involved in transcription.

Initiation of transcription begins with the complex formation at the TATA promoter region. The binding of TF II D factor (Association of TBP with TAF binds to TATA box). This is followed by the joining of TF II A to the complex. After binding TF II A is activated by TBP. The complex is able to protect a region. Binding of TF II B takes place on both upstream and downstream region (– 10, + 10) on DNA. It binds adjacent to TBP and looses its association with DNA. The transcription factor TF II F itself will not bind directly to the complex. It consists of two subunits. The larger subunit is involved in melting the DNA at initiation point. The smaller subunit binds tightly to RNA polymerase. TF II F brings RNA polymerase to the assembly site and finally join the complex. In addition two more factors TF II E and TF II H get attached to the transcription complex. The ATPase and helicase activity of TF II H helps in the movement of RNA polymerase II from the initiation site. Transcription begins with polymerase movement, followed by release of all transcription factors from the DNA.

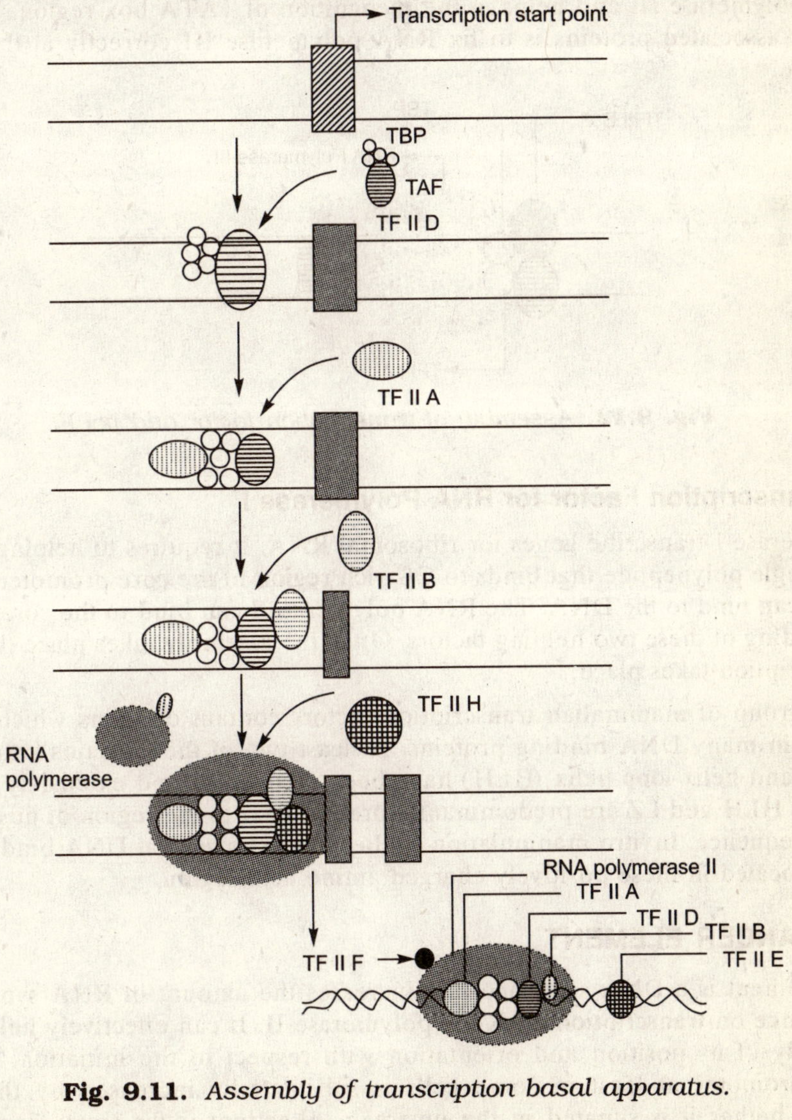

Fig. 9.11. *Assembly of transcription basal apparatus.*

9.16.2 Transcription Factor RNA Polymerase III

RNA polymerase III transcribe tRNA genes like 5S genes. It requires few transcription factors to initiate transcription. Some of the transcription factors like TF II A and TF II B can associate with pol II and polymerase III for the transcription of both protein coding and tRNA genes.

The promoter for 5S tRNA genes are internal (downstream) and are recognized by pol III with the association of transcription factor. The internal promoters recognized by pol III are of two kinds, type one internal promoters and type II internal promoters. Among several transcription factors TF II B is the only factor recognized by RNA pol III to initiate transcription. Factors like TF III A and TF III C are assembly factors, their role is to assist TF III B to a right location on the DNA. TBP could be the subunit of TF III B that interact directly with RNA polymerase III and helps in the recognition of TATA box region. The function of TBP and its associated proteins is to fix RNA polymerase III correctly at the start point of transcription.

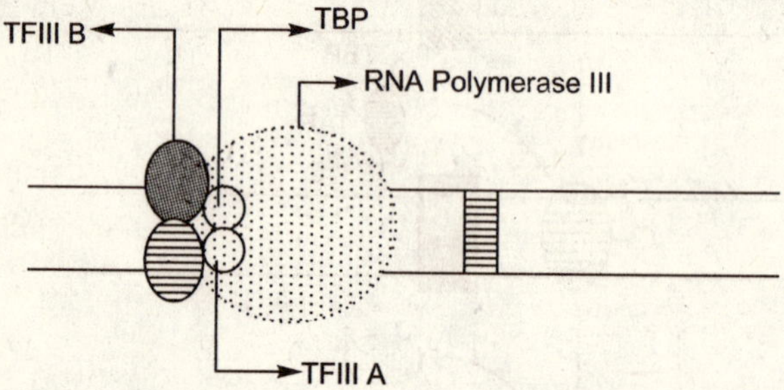

Fig. 9.12. *Assembly of transcription factor and pol III.*

9.16.3 Transcription Factor for RNA Polymerase I

RNA polymerase I transcribe genes for ribosomal RNA. It requires to helping factors, one is UB F 1, a single polypeptide that binds to GC rich region of the core promoter and another is SL 1 which can bind to the DNA. The RNA polymerase I can bind to the core promoter only after the binding of these two helping factors. Once the assembly takes place (UB F, SL 1 and pol I) transcription takes place.

Discrete group of mammalian transcription factors contain domains which are typical of those found in many DNA binding proteins. At least two of the domains known as Leucine Zipper (LZ) and helix-loop-helix (HLH) have been recognised and extensivly studied. These two domains HLH and LZ are predominantly present next to the region of positively charged amino acid sequence. Invitro manipulation studies have shown that DNA binding regions are specifically located in these positively charged amino acid region.

9.17 ENHANCER ELEMENT

Enhancer element is a DNA sequence that increases the amount of RNA synthesis. It has a strong influence on transcription by RNA polymerase II. It can effectively help transcription independently of its position and orientation with respect to the initiation. In otherwords activity of promoter at least in some cells is substantially increased by the presence of enhancers, whether it is situated in the upstream direction or far away from the start site process. The enhancer influence was first detected in mutants of the SV 40 virus. Many viruses including small circular DNA viruses such as polyoma virus and bovine papilloma virus, as well as retrovirus contain enhancer element.

Position of enhancer and its relation to the promoter is not fixed but can vary considerably and it can function in either orientation. An enhancer works prominently when promoter is nearest to it. Several tissue specific transcription is due to the presence of enhancer. Enhancer in co-ordination with promoter act efficiently. A promoter is specifically regulated and presence of nearby enhancer increases efficiency of initiation.

In some cells if promoter is not regulated properly due the lack of specific regulators, then presence of a nearby enhancer activate promoter once itself becomes active. A classical example is the expression of immunoglobulin gene. This gene carries the enhancer within transcriptional unit. The enhancer present in the immunoglobulin gene is active only in the B lymphocytes in which this gene is specifically expressed. Many invitro studies prove that presence of enhancer can influence transcription to several folds. If β-globulin gene is inserted on DNA containing an enhancer, its transcription increases to more than two hundred folds whether it is located several kb in down stream or upstream region.

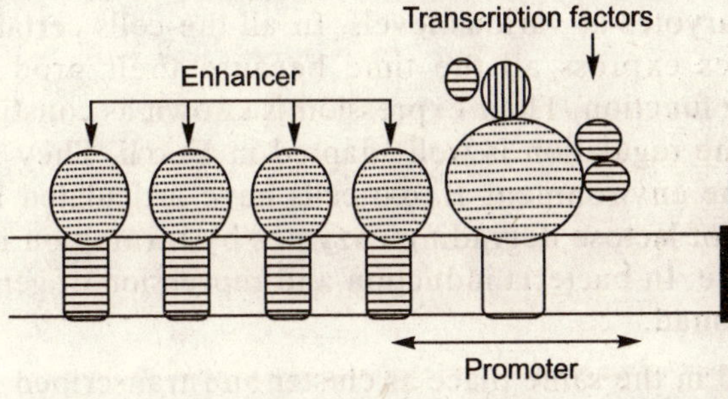

Fig. 9.13. *Presence of enhancer in upstream direction.*

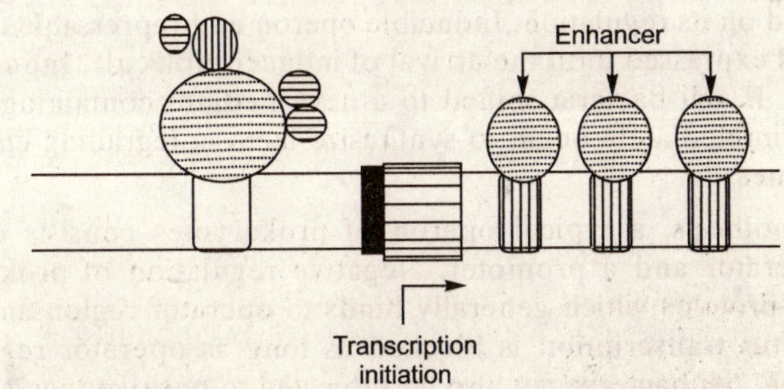

Fig. 9.14. *Enhancer in downstream direction.*

9.17.1 Functions of Enhancers

Enhancer can interact with basal apparatus (Transcription factor assembly) of polymerase II. The exact distinguish between enhancer and promoters are not clear. However, enhancer shows ability to function when it is present at increased distance from the start point. One of the most significant functions of enhancer is enable the assembly of transcription factors in the regions of promoter. Attachment of proteins (transcription factors) at this region i.e., between enhancer and promoter could bring these two into close proximity by protein bridge. This phenomena was confined by invitro experiments. A fragment of DNA contains an enhancer at one end and promoter at the other end. At this position transcription never takes place effectively but the enhancer can activate transcription, when transcription factors connects these two regions.

Work on RNA pol I transcription of the ribosomal genes shows that enhancer also operate in ribosomal DNA transcription. These rRNA upstream sequence stimulate transcription when they are inserted into plasmids in either orientation. They are called ribosomal enhancers.

SUMMARY

Microorganisms like bacteria can quickly adapt to new nutritive environmental conditions due to gene regulation. Bacteria turns on its specific set of genes if products are required immediately. They can switch of their genes when products are not needed. Hence, by regulating their own genes they are able to conserve their energy. Regulation of gene expression takes place in prokaryotes at various levels. In all the cells certain group of genes known as house keeping genes express all the time because their products are required constitutively for house keeping function. Their expression is known as constitutive expression. Induction and repression of gene regulation is well adapted in E. coli. They are able to utilize all types of available food in the environment E. coli cells have articulated in fine regulatory mechanism by which synthesis of lactose degrading enzymes by turning on in the presence of lactose and turn off in its absence. In bacteria induction and repression of gene expression was first described by Jacob and Monad.

In bacteria, genes are associated in the same place as cluster and transcribed as one single unit giving rise to polycistronic mRNA. These clustered super genes are known as operon. Operon contains structural genes, operator gene, regulator gene, and promoter gene. There are two types of operon based on its regulation. Inducible operon and repressable operon. In inducible operon genes are not expressed until the arrival of inducer molecule. Inducers may be lactose, galactose, etc., when E. coli bacteria shifted to a new medium containing lactose as the only carbon source, they immediately begin to synthesize lactose degrading enzyme by regulating the expression of genes.

According to hypothesis, a typical operon of prokaryotes consists of structural genes, regulator genes, operator and a promoter. Negative regulation of prokaryotes is done by regulatory repressor proteins which generally binds to operator region and in the absence of inducer molecule, thus transcription is blocked as long as operator region is occupied by repressor. However, E. coli bacteria can also be subjected to positive regulation under different circumstances by a process known as catabolic repressor in which certain proteins (CAP) are associated with cyclic AMP and are able to bind to the operator promoter region and increase transcription rate. Thus proteins are required for transcription in positive regulation.

Various types of operons have been established based their regulation in response to surrounding environment. They are Lac operon, trp operon, hut operon and gal operon. Their classification is mainly based on the availability of nutrition. In lac operon, lactose degrading enzymes are synthesized when E. coli is grown in lactose containing medium. Lactose acts as an inducer molecule. It will bind to repressor, alter its configuration and then releases from the operator site. Once operator is free, RNA polymerase can recognize this region and transcribe all structural genes.

Tryptophan regulation takes place in a different situation. Trp regulation is an example for repressible operon. For example when E. coli bacteria are grown in presence of tryptophan it immediately turned off its tryptophan biosynthetic enzyme. When tryptophan level is reduced in the medium it begins to synthesise tryptophan biosynthesis enzymes. Trp operon contains 5 genes. Gal operon is regulated according to the nutritional choice of an organism, for example, when E. coli bacteria are grown in presence of both galactose and Lactose in the medium, the organism prefers to use galactose as a energy conservation mechanism.

Hut operon is otherwise called as histidine operon. It acts as inducer molecule. Histidine or hut operon contains four structural genes and two operon one is right handed operon and other is left handed operon. Similarly ara operons consists of three structural genes ara B, A and D (BAD) codes of enzymes to degrade arabinose in the nutritive component.

Eukaryotic system consists of complex array of regulatory mechanism in which only a specific set of genes are expressed in one cell type but not in other. In principle gene expression is controlled at various levels during the process of DNA to RNA and RNA to protein. Hence, regulatory mechanism is active in eukaryotes during the process of transcription, RNA processing, RNA modification and degradation and post transcriptional modification etc.

Davidson and Britton model explains the mechanism of gene regulation among eukaryotic system. In eukaryotic, the regulated genes are distributed in different chromosomes. Most of the genes are in repressed state. Eukaryotic regulatory system consists of several number of sensor sites and integrator genes adjacent to it. Signal molecule can be recognized by sensor genes which inturn pass the message to integrator genes. These are then expressed to produce RNA or protein. RNA products can initiate transcription genes distributed in various chromosomes. Regulatory mechanism has been characterized in other systems of eukaryotes. Well studied systems are galactose regulation in yeast. Yeast contains genes which codes for regulatory proteins for a number of metabolic pathways. These proteins acts as positive and negative level of regulation. Galactose arrival in yeast cell can induce the expression of GAL 4 gene and gal 4 protein in turn play a direct role in the expression of at least three key enzymes distributed on different chromosomes. GAL 1, 7, 10 gene products in turn utilize galactose to produce glucose as final product of the pathway.

The concentration of aromatic amino acids are well regulated in neurospora. When aromatic aminoacids level is increased, neurospora produces some scavenger enzymes viz., quinic acid. These in turn degrade excess aromatic amino acid. Thus, higher concentration of amino acids can be counter productive by excess quinic acid. These in turn acts as a inducer to synthesise scavenger enzyme. When quinic acid is degraded genes for enzymes are automatically switched off.

RNA editing takes place in selected cell types. The mRNA undergo extensive modification by insertion of uridine nucleotide or replacement of one base by uridine nucleotide. They may occupy half of the mature mRNA. There are special RNA molecule known as guide RNA and play an important role in regulation. Regulation can be seen at the level of mRNA transport. The demarcation between nucleus and cytoplasm has greatly enhanced of mRNA contains introns they will not allowed to enter cytoplasm. In addition, signal peptide play an important role in protein targeting. Stability of mRNA is a matter of concern for may cells. Eukaryotic mRNA exhibit longer sheep life when compared to prokaryotic mRNA. The b-globin mRNA is stable for more than 10 hrs. Some of the mRNA life span is less than 30 minutes. Many short mRNA have shown to contain about 50 nucleotides of chromosome rich sequence. Histone mRNA stability is regulated at S phase of cell cycle. Degradation of histone mRNA takes immediately after S phase in cell cycle.

DNA methylation can be linked to gene expression. Amount of DNA methylation has a role in the regulation of gene expression. There are based on three evidences, one is the demonstration of correlation between the level of gene expression and the degree of methylation i.e., low methylation is correlated to high level of gene expression. Similarly high methylation is correlated to low gene expression. Second, methylation is tissue specific. Third, a drug which inhibits methylation results in increased gene expression.

Gene expression is highly regulated by certain class of proteins known as transcription factors. These are categorized into general transcription factors (GTF) and sequence specific transcription factors (SSTF). General transcription factors are indispensable to transcribe all RNA pol II. There are six general transcription factors required to transcribe RNA polymerase dependent genes. These are TF II A, TF II B, TF II D, TF Environment and H, TF II D is an important DNA binding protein. It consists of two subunits one is TATA binding protein component (TBP) and another is TBP associated protein or factor (TAF). Initiation of transcription begins with the assembly of all six transcription factors and RNA polymerases. This complex is known as transcription basal apparatus. There are many transcriptors are available to RNA polymerase I and III. Like TF III A, TF III B and TBP.

Enhancers are unique sequence present on the DNA increases RNA synthesis. It has a strong influence on transcription of RNA polymerase II. They can influence transcription when situated nearer to the transcription site or faraway from it or even when present in upstream or downstream direction of start site.

QUESTIONS

1. Explain the process of gene regulation in prokaryotes.
2. Discuss the process of gene regulation in eukaryotes.
3. Explain with an example how gene expression is regulated in yeast. (BU May 1996)
4. Explain how the levels of tryptophan are regulated in a cell by the process of attenuation. (BU April 1998)
5. Explain the process of Galactose gene regulation in yeast. (BU April 1998)
6. Explain the factors involved in regulation of gene expression in Eukaryotes. (BU April 1998)
7. Describe the regulation of tryptophan operon. (Oct. Nov. 1998)
8. Explain positive regulation of Lac-operon. (BU Nov. 1999)
9. Explain gene expression in yeast. (BU Nov. 1999)
10. How is tryptophan level regulated in E. coli (April 2001)
11. Explain the operon concept of gene regulation with an example. (BU May 2002)

SHORT QUESTIONS

1. Catabolite repression (Nov. 1999)
2. Transcription factors (May 2002)
3. Inducible operon
4. Repressable operon
5. Enhancer sequence
6. DNA methylation
7. Hist. Operon
8. Gal operon
9. Lac operon

10

CHAPTER

Gene Organisation and Expression in Organelle DNA

The chloroplast and mitochondria are the two important cell organelle contains DNA and RNA. They have complete apparatus for protein synthesis. Their resemblance to bacteria suggest that they may be descended from free living bacteria like organism. Deoxyribonucleic acid (DNA) is found in the nucleus of each cell of a plant and animals. However, geneticist have recognized that cytoplasm of all plant and animal cells contain organelle that have DNA of their own. The organelle that found to have contain DNA in plants and chloroplast and mitochondria and animals only mitochondria.

Mitochondria are responsible for producing energy in the form of adenosine triphosphate (ATP). Another organelle chloroplast found in algae and higher plants. The chloroplast contains chlorophyll and related pigments for capturing energy for sunlight and for producing carbohydrate by photosynthesis. During light reaction chloroplast can product ATP energy. Mitochondria and chloroplast was found to be very unique in that they contain not only nucleic acids (DNA and RNA) but also a complete apparatus for the synthesis of proteins. An attractive hypothesis claims that present day chloroplast and mitochondria in the cells were once upon a time free living organism and distantly related to present day bacteria. They were incorporated into larger cells and became endosymbionts during the course of evolution. Extensive studies have been done on how DNA of these organelle influenced their structure and function and to what extent these organelle may be under the control of DNA in the nucleus of a cell.

10.1 FINE STRUCTURE OF CHLOROPLAST

Chloroplast are photosynthetic organelle present in algae and leaves of higher plants. Number of chloroplast per cell, size and shape is highly variable. Higher plant cell contain 40 chloroplast. The shape of chloroplast in algae is ribbon or star shape. Chloroplast in mesophyll cells of leaves of plants exhibit ovoid, discoid and spheroid shape measure 5μm diameter and 3μm tick. Blue green algae and photosynthetic bacteria have no chloroplast. The envelop consist of two membranes of 60-90 Å thick. Outer membrane is separated from inner membrane by intermembrane space of about 10μm.

Each of the internal membrane contains photosynthetic pigments. These seems to be the wall of a flattened sac, sometimes called lamella. At certain region of lamella are organized or stalked to form grana. Each granum contains several short lamella called grana lamellae. The longer lamella that connect one granum to another extend through the chloroplast matrix (Fig. 10.1). The chloroplast matrix is called 'stroma'. It is gel like fluid occupies most of stromal space, which is rich in RNA, protein, lipids and carbohydrates and DNA.

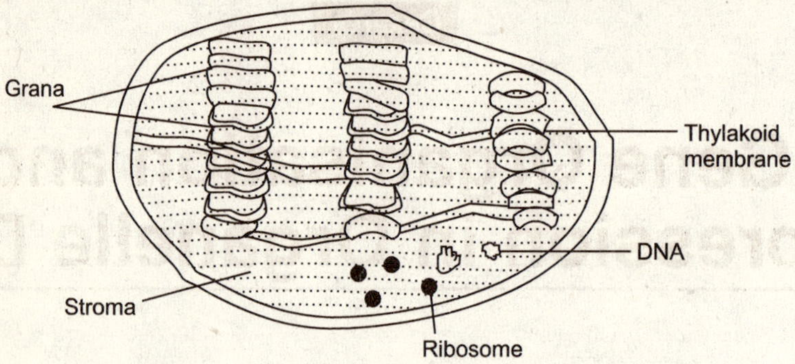

Fig. 10.1. *Fine structure of chloroplast.*

10.1.1 Gene Organization in Chloroplast

Chloroplast DNA is comparatively large, circular in nature. It is denoted as ctDNA. The size of chloroplast DNA is usually 140 kb in higher plants and less than 190kb in lower eukaryotic plants. The chloroplast genome has been completely sequenced in liverworts and tobacco. (Fig. 10.2). The size of ctDNA is between 120-155 kb. Chloroplast contains DNA dependent RNA polymerase this enzyme carry out synthesis of several type of RNA (mRNA, rRNA and tRNA).

The size of the chloroplast genome can be comparable to bacteriophage T_4 (165 kb). There are many copies of DNA in chloroplast i.e., between 20-40 copies per organelle in higher plants. Chloroplast DNA contain genes codes for all ribosomal RNA and transport RNA species needed for protein synthesis. The ribosomes comprises of two ribosomal RNA in addition to major species. It has been found that chloroplast genome code for atleast 50 proteins, including RNA polymerase and some ribosomal proteins.

Chloroplast contains introns. They fall into two classes. One of the class of introns is located in tRNA genes while another class is in protein coding genes. RNA splicing mechanism was found to be well established in chloroplast. One of the key function of chloroplast is in its involvement in photosynthesis. Hence, several of its genes codes for protein are located in thylakoid membrane. Experiments reveals that chloroplast DNA contains 45 genes codes for RNA, 27 genes codes for proteins. These proteins are mainly involved in chloroplast gene expression. While 18 gene are codes for proteins of the thylakoid membrane and another 10 gene products are committed for electron transport.

Chloroplast DNA does not associate with histone protein and thus exist with in the organelle as a nacked fiber. The DNA appears as masses of long strands. However, nacked DNA fibers, devoid of histone and are often exist as ring shaped molecule in a cell having nuclei. Average molecular weight of chloroplast DNA is more difficult to determine. It appears to be some 300 times greater than that of mitochondrial DNA which includes several genes for chloroplast rRNA and genes for proper organization of chloroplast membrane and one or two genes for the synthesis of functional photosynthetic apparatus. The chloroplast ribosomes are probably involved in the synthesis of most of the proteins. Ribosomes of chloroplast shows sedimentation coefficient of 70S Svedberg units i.e., 70S. The ribosomes of cytoplasm exhibit 80S. the 70S ribosomes are made up of 23S and 16S. These RNA are synthesized from chloroplast DNA template. It also contain certain species of tRNA that exhibit properties different from that of cytoplasmic tRNA but these are not necessarily transcribed from chloroplast DNA. Presence of 70S ribosome in chloroplast resemblance to prokaryotic ribosome strengthen the hypothesis of its prokaryotic origin.

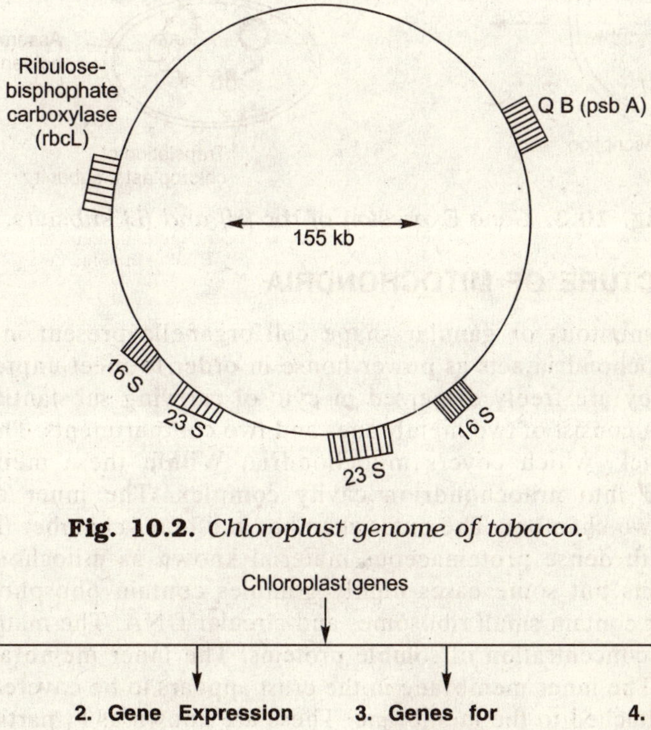

Fig. 10.2. *Chloroplast genome of tobacco.*

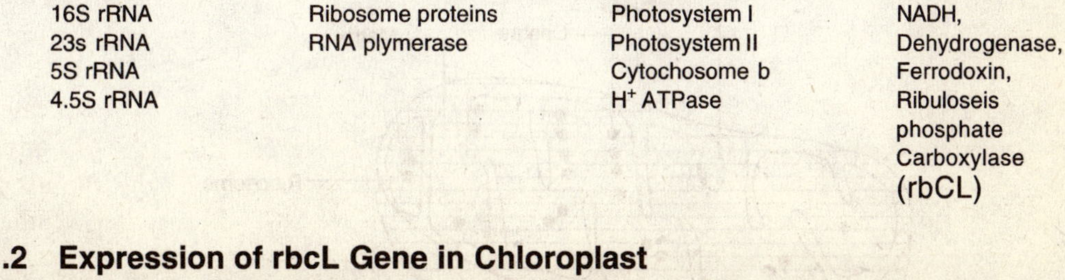

10.1.2 Expression of rbcL Gene in Chloroplast

Expression of ribulose bisphosphate carboxylase large subunit genes (rbcL) have been well established in chloroplast. This enzyme is responsible for fixation of CO_2 in plants as it can binds with high affinity (CO_2). This holoenzyme has eight subunit of which four are smaller subunits and other four are larger subunits. The genes for larger (L) subunits is coded in chloroplast DNA and genes for smaller subunits are coded by nuclear DNA. These nuclear genes when expressed, their mRNA's are translated in cytoplasm. The cytoplasmic 80S ribosomal unit assist in the translation of mRNA. These four smaller subunits are then enters chloroplast with aid of signal peptide. Inside the chloroplast signal peptides are cleaved from smaller subunits and association between larger subunits and smaller subunits takes place to become functional holoenzyme. Association of subunits are aided with the help of chaperon-type assembly proteins. The genes for these chaperon proteins are encoded in nuclear DNA. (Fig. 10.3).

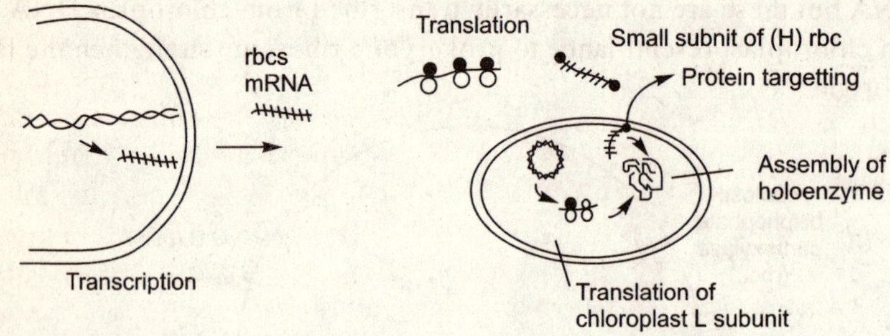

Fig. 10.3. *Gene Expresion of rbc (H) and (L) subsuits.*

10.2 FINE STRUCTURE OF MITOCHONDRIA

Mitochondria is filamentous or ganular shape cell organelle present in the cytoplasm of eukaryotic cells. Mitochondria acts as power house in order to meet unprecedented demand for ATP energy. They are freely dispersed in cytosol carrying substantial amount of ATP energy. Mitochondria consist of two membranes and two compartments. The outer membrane is about 6nm in thick, which covers mitochondria. Within these membrane is a inner membrane projected into mitochondrion cavity complex. The inner membrane divides mitochondrion into two chamber. The outer chamber and inner chamber (Fig. 10.4) the inner chamber is filled with dense proteinaceous material known as mitochondrial matrix. The matrix is homogenous but some cases highly granules contain phospholipids. Within the mitochondrial matrix contain small ribosomes and circular DNA. The matrix appears gel like and contains a high concentration of soluble proteins. The inner membrane infolds to form mitochondrial crest. The inner membrane in the crust appears to be covered by particle of 8.5 nm that have stem attached to the membrane. These are known as F_1 particles. These are the site of ATPase activity involved in coupling reaction.

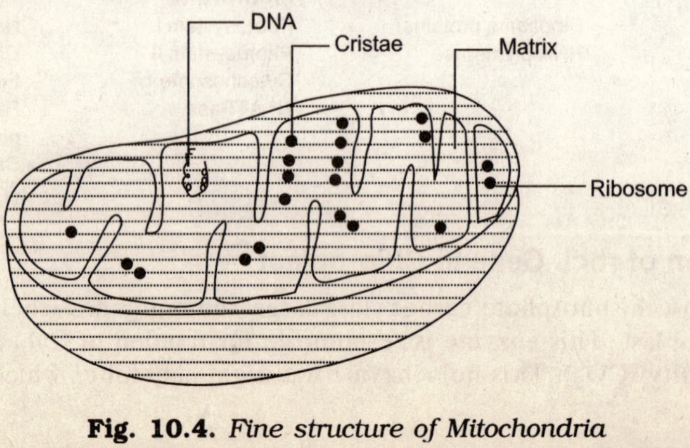

Fig. 10.4. *Fine structure of Mitochondria*

Mitochondria contains circular DNA in their matrix. There is considerable degree of variation could be seen in their numbers. Animal cell contains several hundred mitochondria per cell. These cells have a small mitochondrial genome of 16.5 kb (mammals). The amount of mitochondrial DNA in comparison to nuclear DNA is less than one percent. The plant system exhibit wide range of variations in their mitochondrial genomic size with a minimum of 100 kb in size. High degree of complexity of the plant mitochondrial DNA is due to extensive recombination between the genetic elements.

10.3.1 Yeast Mitochondrial Genome

Mitochondrial genome of yeast and other fungi is about 78 kb long. Large portion of the yeast genomé have already been sequenced. Mitochondrial DNA encodes genes for two ribosomal RNA molecule, complete range of tRNA molecules and mRNA for the synthesis of nine proteins. Mitochondria alone cannot carry out all its genetic activity and other functions. Most of its key functions are controlled by nuclear DNA. The polymerase for the synthesis of mitochondrial DNA and RNA and nearly all the mitochondrial ribosomal proteins and the tRNA synthesis enzymes are encoded by nuclear genes and synthesized in cytoplasmic region. Genes for the synthesis of enzymes for kreb's cycle and electron transport chain is encoded by nuclear genes. All these proteins were made in the cytoplasm, contain special signal peptide sequence for protein targeting it mitochondria.

Mitochondrial rRNA 21S and 15S are smaller than prokaryotic rRNA. The genes for two rRNA are situated on separate portion of the genome. There are many genes which encode for 24 tRNA's. These genes have organized as operon fashion. Some exceptional variation in the genetic code of mitochondria have been determined. It is slightly different from one that is universally used by nuclear genome, including UGA code for tryptophan. According to genetic code by nuclear genome UGA acts as termination codon. The base composition of yeast mtDNA has about 18% G+C ratio.

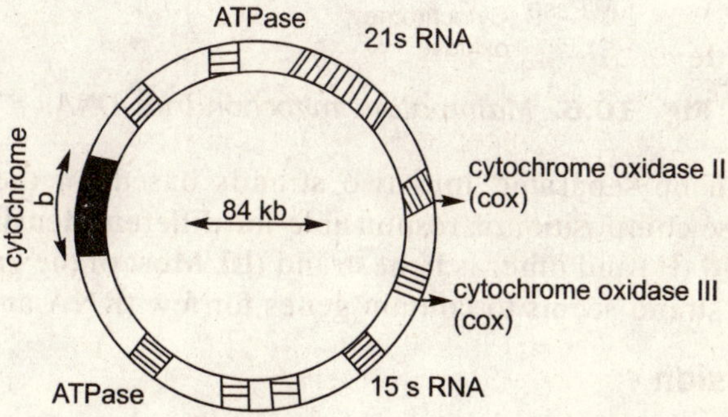

Fig. 10.5. *Yeast mitochondrial DNA.*

Transcription of all the genes present in mitochondrial DNA are initiated by RNA polymerase. It is smaller than those used by transcription in the nucleus. In mitochondria single RNA polymerase are able to transcribe all the genes for the synthesis of tRNA and rRNA and mRNA. Mitochondrial genes promoters contains highly conserved TATA sequence. Several genes encode for mitochondrial proteins are found in the inner mitochondrial membrane. The genes for cytochrome oxidase subunit I and cytochrome β contains five and seven introns respectively (Fig. 10.5) processing of cytochrome β gene products exhibit some interesting features. This gene contains many introns and are removed by splicing process during post transcriptional modification. The enzyme maturase encoded by exon region of the same gene involve in RNA processing. Introns in the cytochrome b, cytochrome oxidase I and also single introns in the 15S rRNA genes are divided into two groups each of which contain unique set of conserved sequence which can assume structure.

10.4 MAMMLIAN MITOCHONDRIAL GENOME

Mitochondrial genome of human, rat and mouse have been sequenced. They are similar in size (16.5 kb). This small mitochondrial genome codes for limited number of proteins. Several mitochondrial proteins are encoded by nuclear genome and are synthesized on cytoplasmic ribosomes. Human mitochondrial DNA has several set of genes like 22 tRNA genes, 2 rRNA genes and 13 protein coding regions. Expression of some of these genes are in clockwise and some are in counter clock wise direction. The proteins which are concerned with respiration are cytochrome b, three subunits of cytochrome oxidase, one of the subunits ATPase and seven subunits of NADH dehydrogenase. Organisation of mammalian mitochondrial genome is highly compact. Introns are not found but some genes are overlapped (Fig. 10.6).

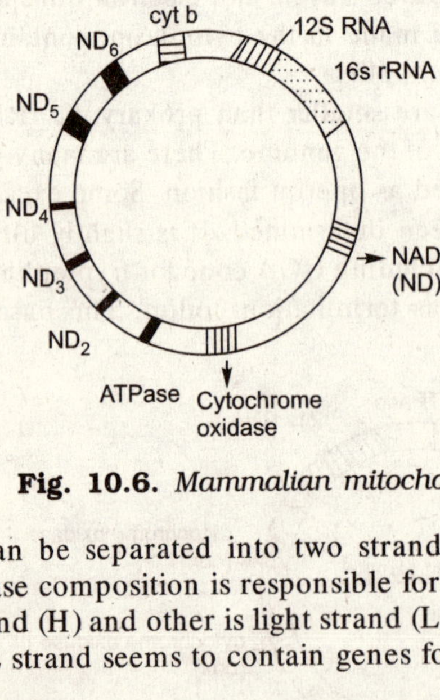

Fig. 10.6. *Mammalian mitochondrial DNA.*

Mitochondrial DNA can be separated into two strands based on their different densities. Overall difference in base composition is responsible for different densities. One of the strand is referred as heavy strand (H) and other is light strand (L). Most of the genes are distributed on heavy strand whereas L strand seems to contain genes for few tRNA and single mRNA.

10.4.1 Gene Expression

The genes in mammalian mitochondrial DNA seems to be transcribed into long precursor RNA molecule. Flanking sequences are virtually absent in mRNA. A number of mRNA are polyadenylated immediately after transcription. Another striking feature is the absence of mRNA capping which is a characteristic features most of other eukaryotic mRNA. The H chain DNA contain more number of genes codes for twelve proteins. The amino acid composition of mitochondrial proteins are hydrophobic. The rRNA in mitochondria are smaller than those made in yeast having 955 and 1550 nucleotide responsible for 12S and 16S RNA respectively.

Mammalian mitochondrial DNA exhibit some difference in their genetic code. In which AUA codes for methionine and can be used as an initiation codon. However, it does not employ N-formyl-methionine for initiation unlike fungal mitochondria. Apart from this codon AUU codes for isoleucine. It is also used as initiation codon. The triplet codon UGA is used for tryptophan rather than termination. In mammalian mitochondrial DNA, AGA and AGG acts as termination codon rather than codons for arginine.

10.5 MITOCHONDRIAL GENOME OF HIGHER PLANTS

Mitochondrial genome of higher plants are different from fungi and animals. They are much larger when compared to the size of fungal mitochondria. Plant mitochondrial DNA is circular molecule of more than 200 kb plant mitochondrial DNA contains number of protein coding genes. They also contains genes for a 5S rRNA resembles the cytosolic ribosomal counterpart.

Table 10.1 Proteins encoded by yeast and mitochondrial genome

Yeast mitochondria	Mammalian mitochondria
Cytochrome oxidase subunits I, II and III	Cytochrome oxidase subunits I, II and III
Cytochrome b	cytochrome b
ATPase subunit 9	ATPase subunit 6, 8
Maturase from introns	

SUMMARY

Apart from nuclear DNA certain cellular organelle also contains DNA of their own. Plant chloroplast contains circular DNA of 190 kb. Nearly 20-40 copies of chloroplast DNA can be seen in a cell. It contains a genes codes for rRNA and tRNA. Genes for larger subunit of rbcL present in the DNA. There are 45 genes coding for RNA, 27 genes for proteins and 10 gene products for electron transport chain. Chloroplast genes contains introns and RNA splicing mechanism. They exhibit 70S ribosomes in assisting protein synthesis more than 100 copies per cell. Mitochondrial DNA is circular with 16.5 kb in size.

They also contain genes codes for rRNA, tRNA and some proteins involved in electron transport chain. The genes for two rRNA are in separate portion of the genome. Mitochondrial genes are transcribed by single RNA polymerase and is smaller than those used by nucleus. Mitochondrial gene promoter contains TATA box for transcription initiation. Based on DNA density, two chains are present in mammalian mtDNA. They are heavy chain (H) and light chain. Human mitochondrial DNA has several set of genes for 22 tRNA and 2 rRNA genes. Most of the genes are distributed on heavy strand whereas L strand seems to contain genes for few tRNA and single mRNA. Mitochondrial genetic code is slightly different from one which is used by universal code. In mitochondrial DNA AUA codes for methionine and universal termination codon AUU codes for isoleucine. Nuclear DNA controls most of the functions of mitochondria.

QUESTIONS

1. Discuss the organisation of organelle DNA.
2. Describe the organisation of chloroplast of mitochondrial DNA.
3. Explain gene expression process in organelle DNA.

SHORT QUESTIONS

1. Mitochondrial genome (BU April 1998; BU Oct. 1998; BU Nov. 1999; 1BU April 2001)
2. Chloroplast DNA (April 2000)

11
CHAPTER

Insertion Elements and Transposons

A comprehensive genetic work has been devoted to the localisation of genes on the chromosomes. It was presumed that the position of the genes are fixed on the chromosomes and do not move to any region. However, in 1940's this presumption was changed after the discovery of movable genetic elements. These are the DNA sequence can change their position. This movable nature of genes was observed by Barbara McClintoc in 1950, working at cold spring harbor laboratory in the maize plant. The movable genes are popularly called as jumping genes or transposons. Transposons can be observed in bacteria, fungi, plants and animals. There are small ranging from 500 to 15,000 bp. They can change their position within the same chromosome or move from one chromosome to another chromosome. Mobile elements can also move from one plasmid to another.

One of the significant feature of mobile elements is that it can alter gene expression. They may inhibit the expression of genes with which they come into close contact or activate genes at the site of its insertion. Transposons often inactivate the genes with which they become attached at any other region. Existence of such mobile genes was thus revealed by their capacity to block functions of other previously characterized genes. They involved in recombination of genetic material.

11.1 INSERTION ELEMENTS AND TRANSPOSONS IN BACTERIA

Bacterial transposons are simple transposons. These are called 'Insertion elements' (IS) or 'Insertion sequences'. The existence of insertion elements in bacteria was based on the discoveryof genetic instability in E. coli result from insertion of alien DNA (Transposon) the bacterial insertion elements or transposons were first to be studied at molecular level. Insertion elements have less than 1500 base pairs devoid of any coding genes for any phenotype character.

11.1.1 Nature of IS Elements

The IS elements occur in chromosomes of bacteria and plasmids. They contain only genes involved in transposition. The genes on the mobile elements are known as transposase genes codes for an enzyme transposase that recognise the ends of the jumping genes and helps in excision and later transposition. The IS elements are compactly organized. On either side of transposase gene is surrounded by short, identical sequence. These terminal sequences are always in inverted repeat sequence. Hence, they are called inverted terminal repeats (IR). Their length range from 9 to 40 base pair. When IS elements insert into chromosome or plasmids. They can create a duplication of the DNA sequence at the site of insertion.

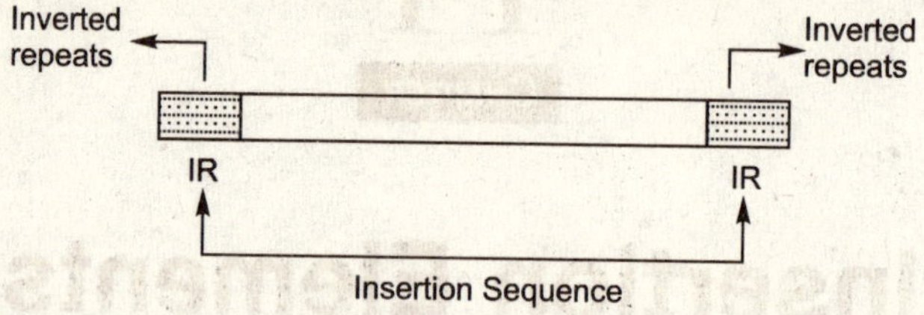

During transposition or after transposition, the sequence of target site DNA undergoes duplication at the site of insertion and this duplicated region is known as direct repeats. This change in sequence was confirmed after investigation. Therefore, only one copy of sequence of target (direct repeat) is present on both the ends of transposon.

In direct repeats altered sequence are inevitable due to transposition. At several site on host DNA insertion of IS elements takes place. The insertion elements was found to contain two coding sequence initiated by AUG and termination by stop codon. Transposable elements in bacteria reveals that each element encode two proteins. Detection of insertion elements can be done by inactivate genes into which they insert and based on the presence of promoter which allow the inactivated gene to transcribe. Therefore adjacent genes are turned on at inserted site. Basically transposable elements acts as a natural agent in reshuffling genetic nature which may result in alteration of structure and function. Several structural observation has been observed in insertion elements. Based on these two generation can be made. In one generation same kind of IS elements can be found on plasmid and chromosome of different bacteria. Second generation is several IS elements can function effectively in bacteria that do not contain IS elements of the same kind. Several copies of IS elements can be seen on bacterial chromosome. E. coli chromosome contain 8 and 5 copies of IS_1 and IS_2 respectively.

11.2 TYPES OF TRANSPOSONS

It contains additional genetic elements like drug resistance may or may not be flanked by IS elements. Transposons are of two types-composite transposons and complex transposons.

11.2.1 Composite Transposons

Which consist of a central region carrying antibiotic resistant genes flanked at both ends by identical copies of an IS elements. As a result of additional antibiotic resistant genes they exhibit drug resistance. These are named as Tn elements. The most well studied composite transposons are Tn_5, Tn_9 and Tn_{10}. Tn_5 shows kanamycin resistance consist of 5,700 base pair and invert repeats shows 1,450 bp at both the end. Tn5 can move (Transpose) from λ phage virus into the chromosome of E. coli.

Tn_9 consist of 2500 base pair with genes for transposase for transposition and genes for chloramphinicole resistance.

Tn_{10} contains antibiotic resistance genes for tetracycline. The number of base pair occur in this transpon is 9300 bp and its invert repeat sequence contains 1,400 base pairs. This can move from plasmid to phage virus.

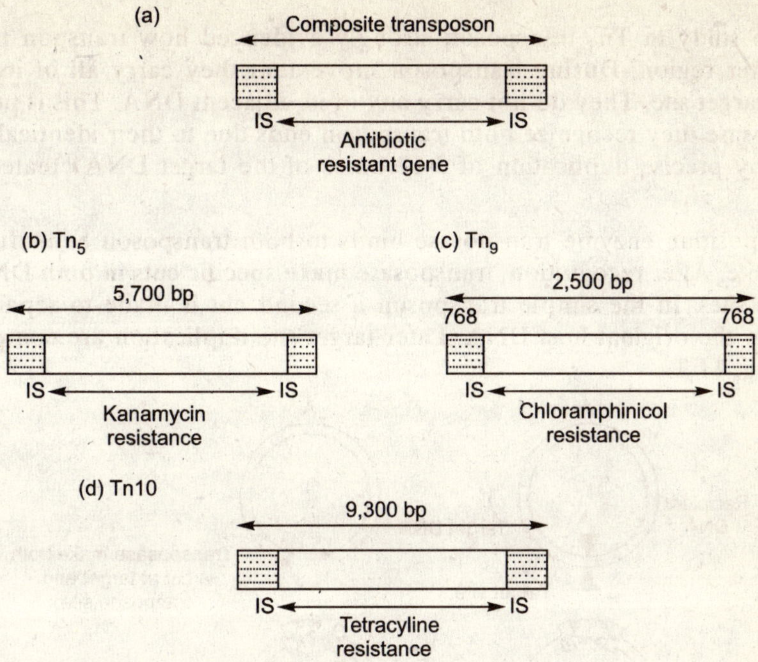

Fig. 11.1. *Types of composite transposons.*

In composite transposons 1S elements can be ne inverted or direct repeat configuration. The ends of IS elements are inverted repeats. The IS elements in this composite transposons are also called as L arm and R arm. Some cases composite transposons are identical in Tn_9 contains direct repeats of IS_1 or Tn 903 contains inverted repeats of 1S 903. (Fig. 11.1).

11.2.2 Complex Transposons

This group includes Tn_1, Tn_2 and Tn_3. These are generally come under Tn A family consist of large elements of approximately 500 bp. They carry genes for transposition and drug resistance. These complex transponsons are larger than IS elements. The Tn_3 transposon having nearly 5000 bp long repeats. These transposon do not move instead it make a copies of its transposon sequence and move to a new location. In Tn_3 genetic organization they contain several genes including gene for transposition. At least three trpA, trpR and bla have been discovered (Fig. 11.2). They codes for transposase, resolvase and betalactamase respectively. The two protein resolvase and transposase helps in transposition and betalactamase confers resistance for the antibiotic ampicillin.

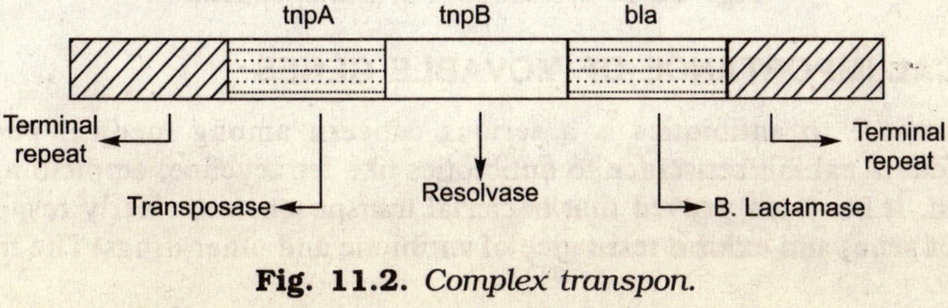

Fig. 11.2. *Complex transpon.*

11.3 MECAHNISM OF BACTERIAL TRANSPOSITION

Comprehensive study in Tn_3 transposon strongly evidenced how transpon move from one region to another region. During transposon movement they carry all of its sequence and inserted at the target site. They do not carry any of its adjacent DNA. This is possible with the enzyme transposase they recognize both transposon ends due to their identical sequence. It is then followed by precise duplication of 3-12 bases of the target DNA created at the site of insertion.

During transposition enzyme transposase binds to both transposon ends. It also recognize the new target site. After recognition, transposase make specific cuts in both DNA site (old site and new target site). In the simple transposon a second cut is made to separate transposon completely from the original host DNA. Later target site duplication are completed by DNA polymerase. (Fig. 11.3)

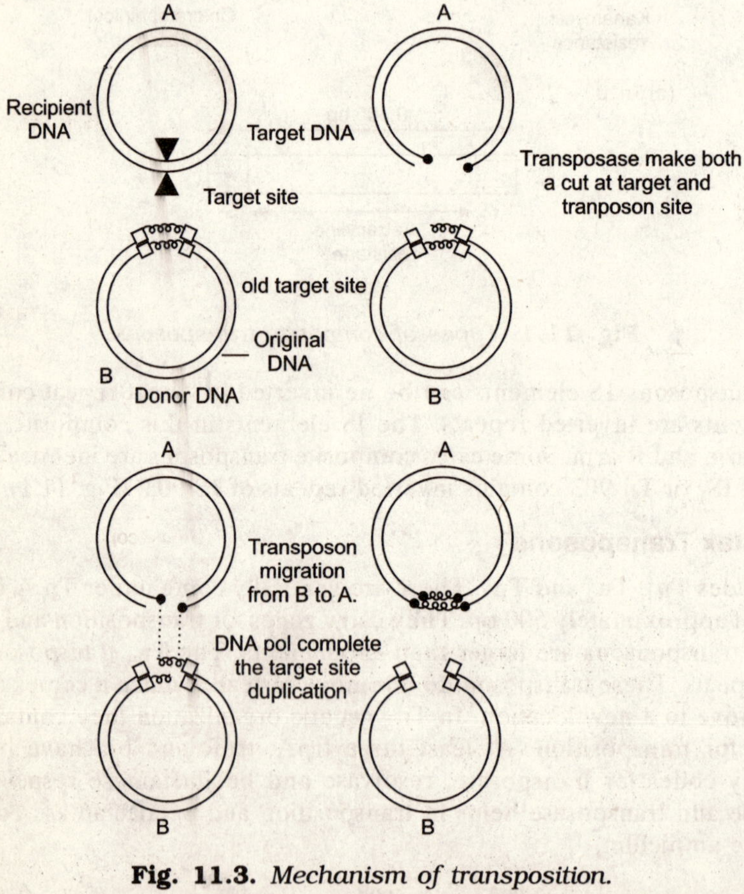

Fig. 11.3. *Mechanism of transposition.*

11.4 MEDICAL IMPORTANCE OF MOVABLE GENES

Acquiring resistance to antibiotics is a serious concern among medical persons. Several pathogenic bacteria exhibit resistance to antibiotics like tetracycline, ampicillin, streptomycin and kanamycin. It has been proved that bacterial transposons are clearly responsible for the transposition of genes and exhibit resistance of antibiotic and other drugs. The transposons are believed to play an important role in the evolution of R plasmid (resistance plasmid) resistance against battery of antibiotics. A simple conjugative R plasmids contain two major DNA component one is the RTF (resistance transfer factor) component which contains the tra genes responsible for conjugative transfer of the plasmid, the second DNA component is R-determinant which carries genes responsible for antibiotic or drug resistance. It is interesting to note that the R-determinant in conjugative plasmids contains identical insertion elements (IS) on either side. The IS elements are probably involved in transposition of R-determinant from one R plasmid to another genetic elements. They can also transfer among species and across generic level. Example in E. coli R plasmid are known to transferred to several genera including salmonella, Haemophiles, Shigella etc. (Fig. 11.4)

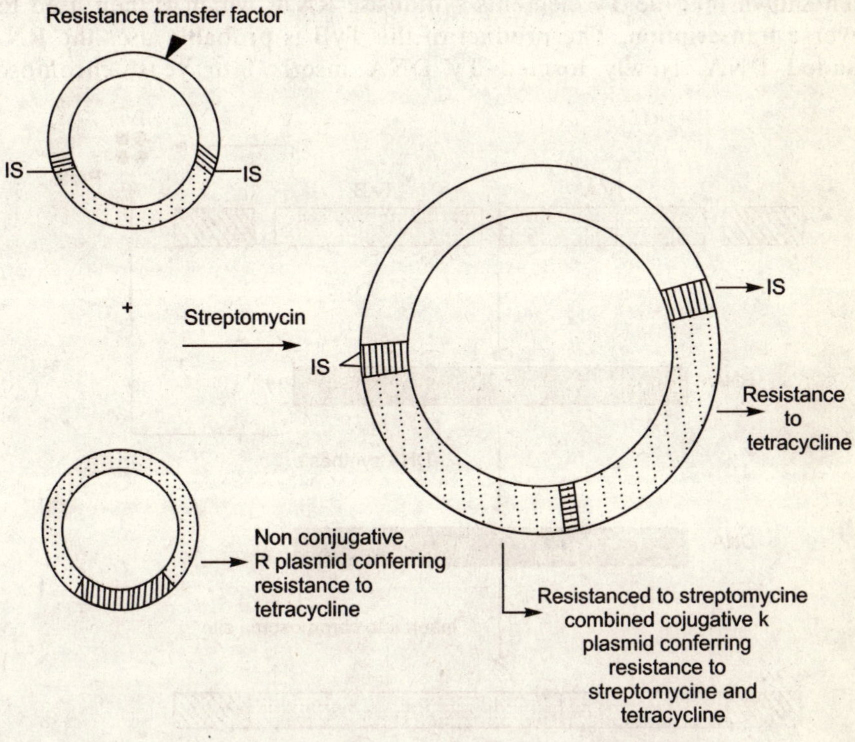

Fig. 11.4. *Mechanism of antibiotic resistance*

11.5 EUKARYOTIC TRANSPOSONS

Extensive work has been done in the field of eukaryotic transposons present in Maize, Drosophilla and yeast.

11.6 YEAST Ty ELEMENTS

The Ty elements of yeast consist of 6000 base pairs. They carry direct repeats of 340 base pairs termed as delta elements (δ). Each delta elements is oriented in the same direction as long termination repeats or LTR. This direct repeat contain a strong promoter for RNA polymerase II leading to transcription of 5.6 kilo base pair mRNA from the left delta elements and transcription of host DNA from the right delta elements. The genetic organization of yeast Ty elements resembles that of eukaryotic retrovirus. They contain two genes TyA and TyB. The products of these two genes from virus like particle inside the yeast cells. It was hypothesized that yeast Ty elements can move from one site to another inside the cell but they fail to move between the cells.

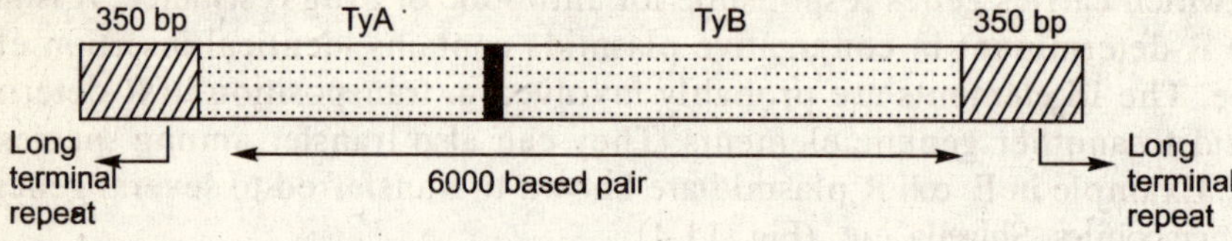

It has been shown that the Ty elements synthesise RNA, which is then used to synthesise DNA by reverse transcription. The product of the TyB is probably uses the RNA synthesis double stranded DNA. Newly formed Ty DNA inserts into yeast chromosomal DNA (Fig. 11.5).

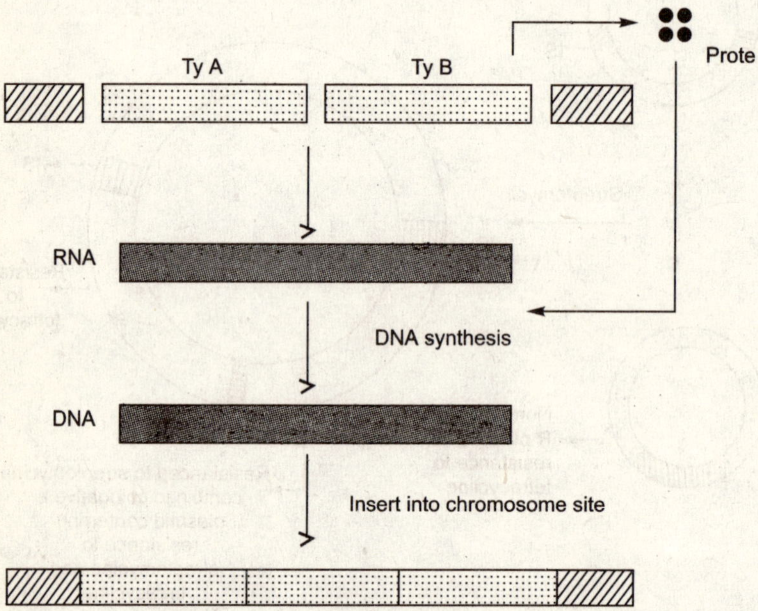

Fig. 11.6. *Mechanism of Ty transposition.*

11.7 TRANSPOSABLE ELEMENTS IN MAIZE

Barbara Maclintock, a nobel prize winner discovered jumping genes in maize. Jumping genes are transposons, known as Ac and Ds elements. Maclintock called those transposons of controlling elements. A_s elements is an transposons with short inverted repeats (IR) flanking sequence encoding a enzyme transposable.

Generally, plant transposable elements have a range of possible activity, all of them associated with possible alterations in genome, i.e., chromosome breakage, chromosomal rearrangement, and gene amplification. Maclintock first characterized D_S as an element that could break chromosomes, thirdly acts as to dissociate the acentric fragment from rest to the chromatid. Chromosome breaking D_S elements have a structure indicative of an element that has insert itself, thus by creating a complex element with three or more ends. It is believed that the inability of the transposable properly rc cognito, mobilize, or repair an attention, excision, of such double D_S element leads to the breakage. A_C and D_S elements are A_C elements carrying deletions in the transposable gene and cannot move on their own. These are two controlling elements are present in ninth chromosomes of cells. The colour pigmentation pattern in maize seeds is due to the presence of allele 'C' on 9th chromosome. A chromosome containing this allele also contain D_S transposon elements can cause break at a particular site on chromosome by another transposable element known as A_C or activator. As a result, non functions of 'C' allele may result in pigmentation synthesis in the maize is 600 bp (Fig. 11.6).

A_s elements excise directly after or during the time that a DNA regulation fork goes through an element, and then to excised elements preferably transposable to a site that has not yet replicated in the S phate. In this process, it creates three copies of A_C i.e., one copy on the replicated donor chromatid and two copies on the replicated target chromatids.

Most transposable elements are usually inactive in single plant or population. It is due to their defective nature. For example, most maize lins have zero or one active A_C element, but two hundred active D_S elements. The inactivation of many transposons in maize is perhaps due to methylation associated process.

Although the chromosome breaking D_S elements are rare, their potential to rearrange the genome is exceptional. Any time when chromosome breaks, this gives abnormal chromosomal (sticky) end that are repaired by telomerrase or will fuse with any a breakage fusion bridge cycle.

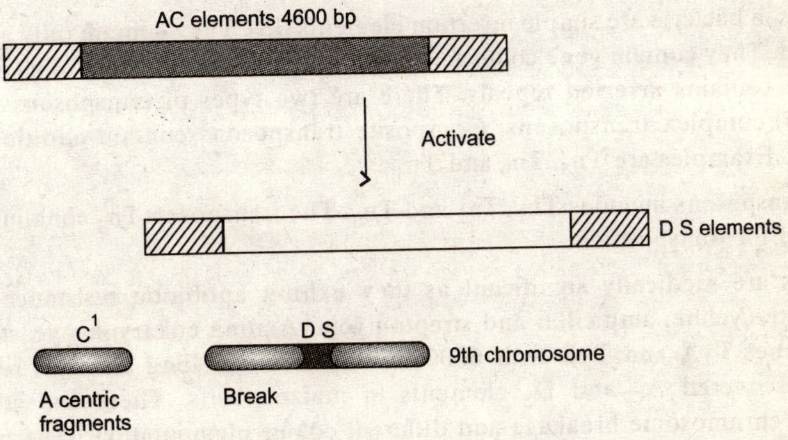

Fig. 11.7. *AC and DC elements of Maize.*

11.8 TRANSPOSONS OF DROSOPHILA

The genome of Drosophila contains number of jumping genes. The amount of transposons in drosophila constitute nearly 15%. One of the largest group of transposon is retrotransposons. These are 5,000 to 15,000 base pairs resembles features of retroviruses and Ty elements of yeast. On either side of the transposon is flanked by long terminal repeat upto several hundred base pair. There are two important retrotransposon family have been studied in detail. One is copia element which is 5000 base pair with flanked region on either side by long terminal repeats. Another is gypsy element contains 7300 base pair with long terminal repeats of 490 base pairs. These transposons are scattered all over the genome occupying different positions in different strains. The mutation is described in drosophila generation is caused by transposons (Fig. 11.8).

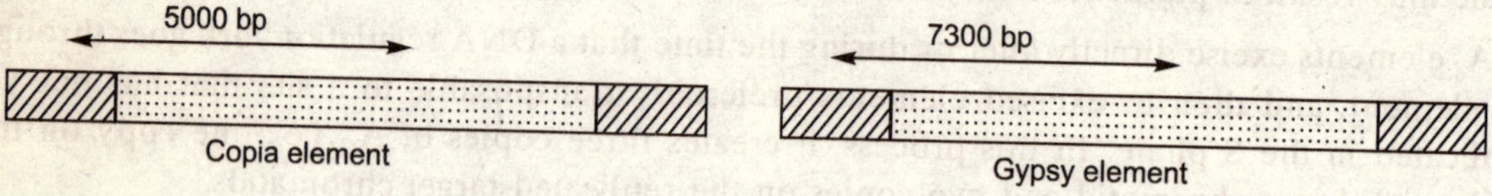

Fig. 11.8 *Drosophila transposons.*

SUMMARY

Transposons are movable genetic elements. They are commonly called as jumping genes. They are present in both prokaryotes and eukaryotes. They can change their position within the same chromosome or move from one chromosome to another or from plasmids to chromosomes or from one plasmid to another. It can alter gene expression or silence the expression of genes or it can activate silenced gene.

Transposons in bacteria are simple insertion elements (IS). They contain only genes involved in transposition. They contain gene codes for transposase helps in excision of its own sequence. On either side contains inverted repeats. There are two types of transposons (a) composite transposons (b) complex transposons. Composite transposons contain additional antibiotic resistant genes. Examples are Tn_5, Tn_9 and Tn_{10}.

Complex transposons includes Tn_1, Tn_2 and Tn_3. The transposon Tn_3 contain at least three genes and gene for transposition.

Transposons are medically significant as they exhibit antibiotic resistance. They show resistance to tetracycline, ampicillin and streptomycin. Among eukaryotes yeast Ty elements contain two genes TyA and TyB with 6000 bp and terminal long terminal repeats (LTR). MaClintock discovered A_C and D_S elements in maize plants. These are jumping genes responsible for chromosome breakage and different colour pigmentation in kernels of maize.

QUESTIONS

1. Jumping genes are ubiquitous in nature. What is the role they play ? (BU May 1996)
2. What are transposons ? Write briefly about transposable elements in any one organism you have studied ? (BU Oct / Nov. 1998)
3. What are Insertion sequences ? Explain their importance.(BU Nov. 1999; BU April 2001)
4. What are transposons ? Discuss the role of transposons in gene expression. (BU April 2000)

SHORT QUESTIONS

1. Transposons (BU April 1998)
2. Insertion sequence (BU Nov. 1999)
3. yeast Ty elements (BU April 2001)
4. Bacterial Transposons
5. Eukaryotic transposons.